Foreword

The project leading to this report is part of a CIRIA programme 'Concrete Techniques Site Operations', and was carried out under contract by Arup Research & Development.

Research Team

M.N. Bussell	–	Arup Research & Development
R. Cather	–	Arup Research & Development

Steering Group

The project was carried out and the report prepared under the guidance of the following Steering Group:

Mr R A McClelland [Chairman]	–	Alfred McAlpine Construction Ltd
Dr P Bamforth	–	Taywood Engineering Ltd
Mr B Cowling	–	Appleby Group
Mr S Crompton	–	Ready Mixed Concrete (UK) Ltd
Dr T A Harrison	–	British Ready Mixed Concrete Association
Mr G G T Masterton	–	Babtie Shaw & Morton
Mr A McGibney	–	Civil And Marine Slag Cement
Dr J B Newman	–	Imperial College of Science, Technology and Medicine
Mr A J Nicklinson	–	Trafalgar House Construction [Major Projects] Ltd.
Mr P L Owens	–	Quality Ash Association
Mr R Roberts	–	Concrete Advisory Service
Mr P Titman	–	Edmund Nuttall
Mr C Turton	–	Design Group Partnership
Dr B W Staynes	–	CIRIA Research Manager

Funding

The project was funded by the Department of the Environment and CIRIA.

Contents

List of Tables

List of Figures

Glossary

Arris — A sharp edge at the intersection of two surfaces.

Cold joint — An unplanned joint at the interface between a new batch of plastic concrete and an earlier batch that has started to harden so that the two batches cannot be merged. Typically such a joint will have reduced bond strength, increased porosity, and may be honeycombed.

Construction joint — A joint between two adjacent concrete pours with no provision for movement. Such a joint is provided for the convenience of the construction process.

Laitance — Cement paste that rises to the top surface of fresh concrete on compaction or is brought to the surface during finishing. In poorly proportioned mixes the laitance may be mostly water, and will dry to leave a soft pasty layer.

Movement joint — A joint formed or induced to allow movement to occur in one or more degrees of freedom.

1 Introduction

This report deals with the provision, specification, and construction of joints in new in-situ concrete construction. It does not deal with repair systems, nor does it cover joints in precast concrete, pavements, bridges or 'massive' civil engineering concrete construction. Both 'discontinuous' and 'continuous' joints are covered.

'Discontinuous' or movement joints are provided to allow movement to take place, usually to prevent damage to the concrete structure and other building components. They may be entirely discontinuous or 'free' movement joints, allowing total freedom of translation and rotation of such joints; or they may be partially discontinuous, allowing free movement along or about certain axes, but restraining or preventing it - and thereby transferring force along others. Types of movement joint are described and illustrated in Section 2.3.

'Continuous' or construction joints, in contrast, arise because it is impractical to pour, compact, and finish all but the smallest structures in a single operation. Construction joints, also commonly called 'daywork' joints, are a practical necessity, and their location and formation must receive careful attention if they are to achieve the required monolithic continuity and not weaken the integrity of the structure.

Both types of joint may have to perform other functions, notably the exclusion of water above and below ground and in water-retaining structures. Discontinuous or movement joints may also have to provide sound and thermal insulation and resistance to the passage of fire and smoke. Durability and appearance are further important factors for both types of joint.

These functional requirements are covered in this report, which gives guidance on the design considerations leading to the provision and specification of joints in in-situ concrete, and describes good practice for their construction. Reference is made to the substantial body of authoritative guidance available, which covers both the philosophy for the provision of structural movement joints and also the design criteria for specifying joints and sealants. This guidance provides background to the specific recommendations of the report.

Recommendations are given for various ways of forming construction joints and research evidence on structural performance is summarised.

Construction joints should always be planned. Pours should be planned and organised to ensure the availability of adequate concrete supplies, labour, equipment, and time to complete an intended volume of pour between planned joints. Accordingly this report does not cover 'ad hoc' construction joints which may be needed when, for example, the supply of concrete breaks down; such a joint will demand quick thinking and action if sound concrete is to be achieved without remedial measures. It would be prudent to consider, in advance, what equipment and facilities by way of reserve, or back-up, would be needed to cope with such eventualities.

This report is intended for use both in the design office and on site. It aims to help structural designers and other members of the design team to make informed decisions about the provision of joints in concrete structures, and to assist those responsible for the construction and supervision of joints.

2 Movement joints: design considerations

2.1 SOURCES OF MOVEMENT

All buildings move. If there is no restraint to movement then it can occur freely, without the development of internal stresses that could lead to damage. In practice some restraint will always be present, even in the smallest building element. The concrete designer is concerned to identify sources of movements, assess their magnitude, and then to consider whether the structure will be damaged if and when movements occur. If so, joints offer one solution for avoiding or controlling such damage. (Other solutions include the use of prestressing to prevent or limit tensile stresses, and the use of additional reinforcement to control cracking.) Failure to provide movement joints where needed may lead to the structure making its own - by cracking.

Sources of movement can be external or internal. External sources include:

- temperature variation
- loading (static and dynamic, including gravity, wind and earthquakes)
- atmospheric humidity changes
- ground movements (settlement, consolidation, shrinkage, heave, etc).

Sources of movement arise also from within the concrete itself. These are principally:

- early-age thermal movement from the rise in concrete temperature during cement hydration and, more significantly, the subsequent drop back toward ambient temperature
- irreversible drying shrinkage
- creep under stress.

CIRIA Technical Note 107[1] discusses these sources of movement. It also provides guidance on estimating movement, limits of 'acceptable' movement, and measures for accommodating movement. This can be read in conjunction with codes of practice for design and construction of concrete structures, namely:

- BS8110: Parts 1 and 2[2],[3] for buildings and structures generally
- BS8007[4] for structures to retain aqueous liquids.

A detailed discussion of the early-age thermal cracking of concrete is given in CIRIA Report 91[5], with guidance on its control. BS8007 necessarily pays particular attention to early-age thermal movement and irreversible drying shrinkage, both of which are of special significance in relation to crack control and watertightness.

2.2 ESTIMATION OF MOVEMENTS AND THEIR SIGNIFICANCE

2.2.1 Introduction

The response of concrete structures to sources of movement is dependent on a variety of factors which, considered together, make estimating movements an approximation at best. Nevertheless it is possible to estimate orders of magnitude as a prelude to considering what action (if any) needs to be taken, before reviewing the possible ways of achieving this to produce satisfactory performance.

Data for estimating thermal and moisture movements and stresses in many constructional materials have been published in three BRE Digests, 227-229[6]. Quantitative data relevant to concrete were reproduced in CIRIA Technical Note 107, and are quoted below with supplementary material from other authoritative sources.

It must be stressed that the following is concerned with estimating the movement of the concrete structure. In buildings and other structures, the issue of movement of other constructional materials, and their possible interaction with movement of the concrete structure, needs to be considered. For example, vertical movement joints are recommended at spacings of typically 6 to 15m for masonry walls in BS5628: Part 3[7], depending on the masonry units e.g. concrete blocks, calcium silicate or clay bricks. This is based on experience of cracking in such walls built with present-day cement-based mortars. This may have no significant interaction with movement of concrete elements in the same structure, except that it is desirable for any movement joint in the concrete to pass also through the masonry and finishes. If this is not done the components bridging the joint will almost certainly be cracked by movements of the concrete. On the other hand, non-loadbearing masonry walls supported by a framed concrete structure may need provision for horizontal head joints to accommodate differential movements between the masonry, the concrete slabs and beams over (which will deflect when loaded, by shrinkage, and by creep) and also the adjacent concrete columns and/or walls (which will shorten elastically, by shrinkage, and by creep when loaded). CIRIA Technical Note 107 and BS5628: Part 3 give advice on such issues.

2.2.2 Temperature variation

BRE Digests 227-229[6] give examples of service temperature ranges for concrete in the UK (reproduced from CIRIA Technical Note 107) when fully exposed and when used internally. These are reproduced here in Table 1.

Table 1 *Concrete temperature ranges in service*

Location and condition	Minimum (°C)	Maximum (°C)	Range (°C)
Fully exposed concrete structural members:			
Light colour	-20	45	65
Dark colour	-20	60	80
Internal:			
Normal use	10	30	20
Empty/out of use	-5	35	40

Technical Note 107 points out that variation of temperature from its installation or datum value is often more important, the latter usually being somewhere within the range 5-25°C.

Typical values of coefficients of linear thermal expansion for concretes are given in Technical Note 107 and are reproduced here in Table 2.

Table 2 *Thermal expansion coefficients of concrete*

Material	Coefficient of linear thermal expansion α (per °C $\times$ 10^{-6})
Dense aggregate concrete:	
Gravel aggregate	12-14
Crushed rock (not limestone)	10-13
Limestone	7-8
Lightweight aggregate concrete:	
Medium lightweight	8-12
Ultra lightweight	6-8
(expanded vermiculite and perlite)	
Aerated concrete	8
Steel-fibre-reinforced concrete	5-14

Similar, but not identical, values are given in BS8110: Part 2 and CIRIA Report 91. These values, it must be stressed, are typical: variation of material sources may result in wider ranges.

To put some order of magnitude to these temperature-induced movements, consider a concrete structure 60m long built with dense gravel aggregate concrete.

If the structure was built at around 15°C ambient temperature, then the overall expansion or contraction can be calculated as (α $\times$ overall length $\times$ temperature change).

In the case of an external member of light colour, the overall contraction between 15°C and -20°C is (12 to 14) $\times$ 10^{-6} $\times$ 60 000 $\times$ (15-(-20))mm = 25.2mm to 29.4mm.

In a symmetrical structure supported on columns (Block A in Figure 1), the contraction at one end would be one-half of this, 12.6 to 14.7mm.

Such a contraction considered as horizontal movement over a storey-height of say 3m between a thermally-stable ground floor slab and an exposed first floor slab would represent a column 'drift' (sidesway divided by height) of (12.6 to 14.7)/3000 = 1 in (204 to 238), which CIRIA Technical Note 107 suggests may be unacceptably large in serviceability terms. Were the building to be stabilised by longitudinally stiff core walls at each end (Block B in Figure 1) then such drift would certainly induce substantial restraint forces in these walls.

It must be acknowledged that such an example calculation perhaps represents a 'worst credible' situation, with externally fully exposed structural members and a sustained period of extreme low temperature of sufficient duration to allow the entire structure at first floor to cool to such an extent. It thus also neglects any thermal relief from building services or insulation. Nevertheless the order of magnitude of such estimated movement is sufficient for the provision of movement joints to be considered.

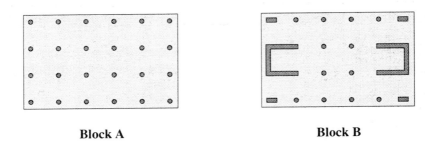

Block A   Block B

Figure 1 *Typical floor plans of buildings subject to thermal movements*

2.2.3 Early-age thermal movement

Early-age thermal movement comes about from the rise in concrete temperature during cement hydration and its subsequent fall back towards ambient temperature. The fall is more significant, as it leads to contraction of the 'green', newly-hardened concrete of low tensile strength, with the consequent risk of cracking. This is especially important in watertight and water-retaining construction. The issue is discussed with authority and at length in CIRIA Report 91[5]. Summary guidance is given in BS8007[4], and in BS8110: Part 2[3]. These sources contain guidance on how to minimise early-age thermal cracking by firstly controlling the temperature gain in the concrete, and secondly by keeping the resultant cooling contraction strains within the tensile strain capacity of the young concrete by judicious construction planning to take account of restraint from adjacent hardened concrete. The problem is of particular concern with large-volume concrete pours, where large heat gains cannot easily be dissipated[9],[34].

The risk of cracking can be mitigated as much by design and construction planning as by the introduction of joints. Both approaches need to be considered, especially in large-volume pours where joints alone will not prevent damage from excessive temperature differentials between the core and the surface. Here, other measures such as cooling of materials and use of insulation will be of greater benefit in reducing cracking than provision of movement joints.

2.2.4 Atmospheric humidity changes

Changes in atmospheric humidity affect structural materials. This is additional to the drying shrinkage of concrete after hardening (see Section 2.2.5). CIRIA Technical Note 107 quotes the typical values given in Table 3, which are reversible. Drying produces shrinkage; moistening produces expansion.

Table 3 *Reversible moisture movement in concrete*

Material	Reversible moisture movement (%)
Dense aggregate concrete:	
Gravel aggregate	0.02-0.06
Crushed rock (not limestone)	0.03-0.10
Limestone	0.02-0.03
Lightweight aggregate concrete:	
Medium lightweight	0.03-0.06
Ultra lightweight	0.10-0.20
(expanded vermiculite and perlite)	
Aerated concrete	0.02-0.03
Steel-fibre-reinforced concrete	0.02-0.06

Except for the fibre-reinforced concrete, these values apply for unreinforced concrete, as it is accepted that reinforcement can provide restraint and hence reduce movements to some extent[3].

Considering again the example of a 60m long first floor slab of dense gravel aggregate concrete, the reversible moisture movement (between the two theoretical extreme conditions of oven dry and fully saturated) would be in the range of (0.0002 to 0.0006) × 60 000mm = 12 to 36mm.

In practice such extremes are not achieved even fleetingly in a 'real' UK climate. BS8110: Part 2[3] suggests that credible typical internal and external relative humidity values of 40% and 70% respectively may be appropriate for concrete. The 30% difference between these (instead of 100% as calculated above) leads to estimates of reversible moisture movement of 3.6 to 10.8mm, and even these are based on complete transition from 'internal' to 'external' humidity or vice versa. This is improbable.

Experience indicates that atmospheric humidity changes do not on their own warrant movement joints in concrete.

2.2.5 Irreversible drying shrinkage

Concrete shrinks irreversibly as it slowly loses moisture after hardening. CIRIA Technical Note 107 quotes typical values for this as shown in Table 4.

Table 4 *Irreversible drying shrinkage in concrete*

Material	Irreversible drying shrinkage (%)
Dense aggregate concrete:	
Gravel aggregate	0.03-0.08
Crushed rock (not limestone)	0.03-0.08
Limestone	0.03-0.04
Lightweight aggregate concrete:	
Medium lightweight	0.03-0.09
Ultra-lightweight (expanded vermiculite and perlite)	0.20-0.40
Aerated concrete	0.07-0.09
Steel-fibre-reinforced concrete	0.03-0.06

These values are generally derived from laboratory specimens allowed to undergo free contraction. Except for the fibre-reinforced concrete, they apply for unreinforced concrete, as it is recognised that reinforcement can provide restraint and hence reduce movements to some extent[3].

Before these figures are applied directly to calculating structural movements or stresses, two factors need to be considered. One is the effect of creep, which reduces the actual tensile stresses due to shrinkage (see Section 2.2.8 below). The second is that, for most floors of a structure, drying shrinkage is a progressive process whose effect moves up the structure as it is built, producing a 'ripple' of contraction. This will substantially reduce differential shrinkage movements. The resulting stresses can usually be accommodated by designed and nominal reinforcement.

Special cases needing particular attention to design for shrinkage effects are large ground floor slabs, water-containing/retaining structures, and elements linking sections with differing stiffness. Where practical, ground-bearing slabs can be constructed on a slip membrane to allow contraction to occur, aided by sufficient reinforcement to control shrinkage cracking[11]. The design of water-retaining structures must give particular attention to crack control; guidance is given in BS8007[4]. Special care will be needed where the stiffness of the structure against shrinkage movements is locally reduced. An example is a narrow 'bridging' section between larger buildings (see Figure 6) - for which at least one movement joint should be provided to minimise the risk of significant cracking. Buildings with substantial longitudinally stiff walls or columns at each end, such as Block B in Figure 1, are likewise vulnerable to shrinkage effects as well as to temperature variations (see Section 2.2.2).

2.2.6 Loading effects

Loading effects come from the building self-weight, user loads, wind, earthquakes, and also incidental secondary effects such as out-of-plumb or lack of column or wall concentricity between adjacent levels. They are generally of a static nature, although dynamic effects can be significant. These arise from wind, earthquakes, vibration of installed equipment, etc.

Estimation of movements due to loading is generally covered in codes of practice for concrete; such movements do not normally result in the need for structural joints. They may, however, give rise to the need for separation joints between non-loadbearing masonry walls and structure, as mentioned in Section 2.2.1. An exception where movements due to loading can cause damage is the case of seismic action. Joints are often introduced into buildings with unsymmetrical structures (see, for example, Figure 7) to minimise eccentricity between the building's centre of mass and the structure's shear centre.

2.2.7 Ground movement

Ground movement, or rather differential ground movement, is a principal justification for movement joints in concrete structures.

Ground movement occurs most commonly as a result of:

• settlement under loading
• consolidation of compressible soils (e.g. clay) due to changes in loading or moisture content

- subsidence due to mining or tunnelling
- changes in groundwater level after construction.

Movement often occurs both vertically and horizontally at foundation level, so that buildings may be at risk from differential vertical movement causing racking distortion and also from horizontal movement causing tensile or compressive strain.

A 'classic' location is between two blocks of different height, particularly where ground conditions warrant different foundation schemes. Figure 2 shows a tower block with piled foundations, abutting a low-rise block on spread footings or a raft. The site is underlain by clay, for which the foundation solutions chosen are both logical and economical. It is recognised that the settlement characteristics of piled and spread or rafted foundations are different[8] and so it is logical to separate the two blocks by a joint allowing independent movement.

The reliable estimation of ground movement is no easier than the estimation of movements in concrete. Ref. (8) draws together authoritative guidance.

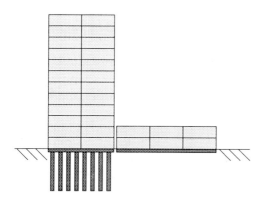

Figure 2 *Movement joint location determined by ground and foundation considerations*

2.2.8 Creep

Creep is a time-dependent phenomenon whereby stressed concrete gradually deforms further. Thus a stress applied and sustained will cause creep. 'Young' concrete creeps more than old, and concrete subjected to sustained stress will creep more than concrete only briefly stressed. Under normal service loads creep movement will reach a limiting magnitude, but further creep would occur if loading were subsequently increased.

Where the stressing effect is maintained, creep simply increases strains. Thus a beam or slab under gravity loading will deflect further due to creep, eventually deflecting to a limiting extent. But when the stressing effect is internal, e.g. due to shrinkage, then relief of stress can occur, with significant reduction in residual strain. Concrete Society Technical Report 22[10] reported long-term residual tensile strains due to shrinkage, with creep relief, of the order of 30 microstrain (a strain of 30 x 10^{-6})

compared with short-term small-scale laboratory results - without creep relief as they were unrestrained - of 300 microstrain, a 90% reduction.

BS8110: Part 2[3] gives guidance on estimating notional creep enhancement of initial deflections, but makes no allowance for creep relief as outlined above.

Creep in itself will not warrant provision of joints in orthodox structures. However, it may be significant in special circumstances, for example a tall building with a slipformed concrete core and perimeter steel columns coupled by occasional outrigger steel trusses to resist wind forces (see Figure 3). Vertical shortening of the core due to creep (and also shrinkage) after construction could induce damaging stresses in both trusses and perimeter columns unless provision was made for controlled release and re-tightening of steel joints at intervals.

Creep effects will also compound the need for joints between a concrete structure and non-structural components such as infill masonry walls.

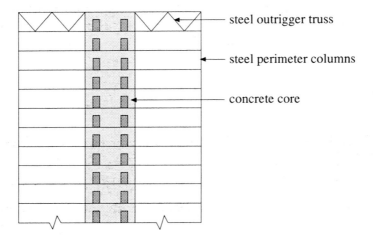

Figure 3 *A structure where creep effects may warrant provision for movement*

2.3 TYPES OF MOVEMENT JOINT

From the discussion in Section 2.2, it will be seen that movement joints should essentially be considered for coping with:

• temperature variation movements
• early-age thermal movement
• irreversible drying shrinkage
• ground movement.

Alternative types of movement joint are now described.

2.3.1 The free movement (or isolation) joint

As its name implies, this joint allows free translation and rotation in all directions see Figure 4(a). It is most commonly used in the following circumstances:

- In structures above ground level (particularly in roof slabs and exposed soffit slabs) to accommodate temperature variation movements. Temperature variations are both of smaller magnitude and slower to occur below ground level, due to the 'heat sink' effect of the subsoil and reduced exposure to climatic extremes.

- In structures generally, to accommodate differential ground movements, especially when adjacent elements or blocks exert different levels of bearing pressure and/or have foundations with different settlement characteristics; such free movement joints are also known as settlement or isolation joints.

2.3.2 The free contraction joint

This type of joint is formed, and has no initial gap. It is for use where movement will lead to opening of the joint only (see Figure 4(b)). It is most commonly used in water-containing structures to deal with early-age thermal movements and irreversible drying shrinkage, where no load transfer or equalising of deflection in the plane of the joint is required. Such a joint may be applicable in circumstances where some expansion may occur - for example due to temperature rise - but only when this follows, and is of lesser magnitude, than initial contraction.

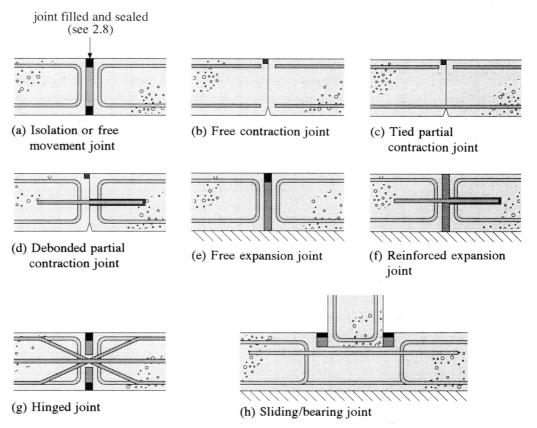

Figure 4 *Types of movement joint*

2.3.3 The partial contraction joint

In this type of contraction joint, reinforcement continues across the butt joint in the concrete although it is either reduced in section, or debonded, so that the joint will indeed be able to serve as a plane of weakness at which contraction can occur[53]. Steel is provided to ensure that shear loads can be transferred across the joint and/or when equalising of deflection in the plane of the joint is required. Such joints are provided to deal with early-age thermal movement and irreversible drying shrinkage in water-containing structures, retaining walls, and large ground-bearing slabs. There are two basic variants - the tied and the debonded joint (see Figures 4(c) and 4(d)). They may be formed by placing concrete either side of the joint in two pours, or alternatively the concrete is placed in a single pour. In the latter case separation at the joint is achieved by use of a crack-inducing strip and/or sawing a groove in the concrete surface.

The tied contraction joint

The tied contraction joint has a reduced area of reinforcement across the joint. This assists shear transfer and prevents overall free opening of the joint, while permitting relief of early-age thermal movements and shrinkage on the surface. (This reduces the possibility of unwelcome surface cracking from such causes occurring elsewhere.)

This type of joint is much used between adjacent pours in water-containing structures and large ground-bearing slabs.

The debonded contraction joint

In this variant, some reinforcement is provided across the joint as for the tied contraction joint. It is debonded on one side of the joint so that unrestrained contraction can occur across the full thickness of the section to allow early-age thermal movements and shrinkage to occur. This type of joint is used less often than tied contraction joints. It is most commonly used in large ground-bearing floor slabs, roads, and hard-standing. The reinforcement is usually provided in the form of dowel bars.

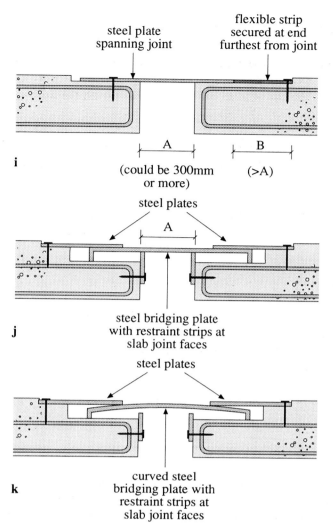

i (could be 300mm or more) (>A)

j steel bridging plate with restraint strips at slab joint faces

k curved steel bridging plate with restraint strips at slab joint faces

Figure 4 *(continued) Types of movement joint - seismic/open joints*

2.3.4 The expansion joint

This term is commonly, but loosely, used to describe any formed gapped joint, whatever its structural role and irrespective of whether opening or closing movements are expected.

Such a joint is intended to allow expansive movement to occur freely. It may be free (Figure 4(e)) - in which case it is essentially a free movement or isolation joint - or have reinforcement (Figure 4(f)) to transfer shear and equalise deflections. The reinforcement, if present, must be debonded to allow free axial movement. This type of joint is used for elements exposed to significant temperature variation (notably solar gain) such as roof slabs, footbridges, and ground-bearing slabs outdoors.

2.3.5 The hinged joint

The hinged joint (Figure 4(g)) is more common in bridges, particularly arches, than in building structures. It has a narrow concrete throat with concentrated reinforcement to allow rotational freedom (minimising moment transfer), while providing shear and axial load transfer and equalising deflections across the joint. It has been used in a few structures to provide effectively 'pinned' concrete connections, for example to minimise potentially damaging horizontal shear forces in the lowest lift of columns carrying floors down to a progressively-post-tensioned long-span transfer girder (Figure 5).

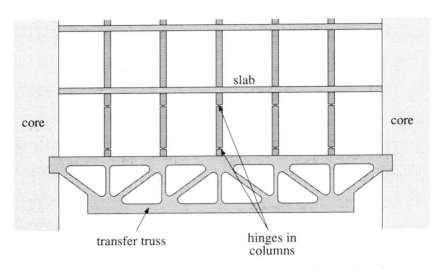

Figure 5 *Hinged joints to relieve potentially damaging forces*

2.3.6 Sliding joints and bearings

Sliding joints and bearings are generally used in precast concrete structures and in larger civil engineering structures such as bridges, although sliding joints are also found at the wall-floor and wall-roof junctions in water-retaining structures. Such joints are similar to free movement joints except that loads are carried from one element to the other by bearing. This may be achieved either directly across smooth concrete surfaces or (preferably, to minimise friction) by purpose-made bearings or membranes designed to allow the required freedom of movement while still transmitting loads. Such bearings, suitably designed, can also provide acoustic and vibration isolation (e.g. for a hospital built next to a railway line) and base isolation to protect structures in seismic zones.

A common example of a sliding joint is the slip membrane provided under a ground-bearing slab to allow shrinkage contraction. This is discussed in Deacon[11].

Figure 4(h) shows a typical sliding joint at a floor-wall junction in a water-retaining structure.

2.3.7 Seismic joints

Earthquakes can cause large movements to occur in buildings and other structures. This calls for particular attention where open movement joints have been provided; these joints will have to be of sufficient width to accommodate the cumulative anticipated movement across the joints. This may be 100mm or more, which is well beyond the capability of any orthodox joint sealant or gasket. Choosing a suitable bridging material for such joints may be complicated by the need to accommodate pedestrian or vehicular traffic.

One approach is to provide steel plates, either fixed on one side of the joints or loose but 'captive' within retaining strips. Figures 4(i), 4(j) and 4(k) show three variations.

2.4 OTHER FUNCTIONAL REQUIREMENTS OF A MOVEMENT JOINT

The philosophy of joint design in building construction is thoroughly covered in BS6093[12] which includes an extensive list of possible functions of a joint. Of these, the most generally relevant non-structural requirements are:

- durability
- resistance to water passage (as ice, liquid, and vapour)
- sound insulation
- thermal insulation
- resistance to fire
- appearance
- accessibility for inspection and maintenance.

Other considerations may of course apply in particular situations: the above list is certainly not exhaustive.

These common requirements are discussed below.

2.4.1 Durability

Ideally a joint should have the same design life as the structure, but in practice this is difficult to achieve, as the components of any joint, other than an unsealed free movement joint, are subjected to exposure and to cyclical movements and forces. These components therefore must be selected with care, balancing first cost against the consequences of joint failure and the practicability of access for inspection and maintenance (including replacement if necessary).

It will often be wise, despite pressures on cost grounds, to adopt what is initially an expensive but durable and robust design for a joint to ensure that it will last the expected lifetime of the structure. This will certainly be the case for components whose replacement will involve substantial disruption, such as dowel bars transmitting load across the joint between two panels of an industrial floor. It is essential where failure of components could be catastrophic, as in dowel bars supporting an exposed stair landing cast after the stair walls. In such cases the use of stainless steel or non-ferrous bars should be considered, despite their greater cost than plain or lightly-protected carbon steel.

BS7543[13] introduces frameworks and concepts upon which the issues of durability for variable design life can be set.

2.4.2 Resistance to water passage

Water leakage and ingress are of major concern in building performance, and joints in a concrete structure must be carefully designed and detailed to achieve satisfactory performance[4],[17],[18].

Exposed joints in concrete above ground level will usually be subject to atmospheric humidity and precipitation in liquid and solid forms (rain, snow, hail, etc), as well as incidental wetting as from a hose or leaking service pipe. The normal solution is to seal the joint using materials able to accommodate the anticipated joint movements (see Section 2.8). Joint design for such conditions is discussed in BS6093[12].

Joints in concrete at or below ground level will usually be subject to groundwater in both liquid and vapour forms. It may be necessary to consider impurities and pollutants in the ground, and their potential effect on contents, occupants, building use, and construction materials. It may be necessary also to consider groundwater levels over the expected building lifespan, taking account of evidence for rising levels in urban areas[14] [14a].

Criteria for moisture resistance in buildings as this affects health and safety are laid down in the Building Regulations[15], and Approved Document C[16] indicates ways of complying with the requirements. BS8102[17] is cited, and is a general source of guidance for protection of concrete structures against water from the ground. Design solutions often include waterstops to provide a physical barrier to ground moisture, instead of or in addition to sealants (see Sections 2.7-2.8).

Joints in ground-bearing slabs will often be subject to moisture from both above and below, and usually need to be sealed to resist this moisture as well as pollutants (including solvents and abrasives, particularly in industrial use). Design guidance is available[11], [12], [53].

Joints in structures retaining aqueous liquids must of course be sealed to avoid undue leakage outwards and also contamination from leakage inwards. In addition, attention must be given to ensuring that sealant materials will have no adverse effect on potable water and be mutually compatible with other chemicals in the retained liquid.

2.4.3 Sound insulation

An open joint will intercept structure-borne sound but provide a path for airborne sound transmission. This is sometimes exploited by acoustic designers, for example in theatres. On the other hand, some form of external or internal insulation may be required. BS8233[19] gives guidance on assessment and solutions. It is important that any sound insulant provided across or within the joint can accommodate anticipated movements (and is compatible with other performance requirements of the joint). The insulant will therefore have to be flexible.

2.4.4 Thermal insulation

Similar considerations apply as for sound insulation. Approved Document L of the Building Regulations 1991[20] and a BRE publication on avoiding risks in thermal insulation[21] offer guidance.

2.4.5 Resistance to fire

The fire resistance of joints in buildings must comply with Building Regulation requirements[15]. Approved Document B[22] indicates ways of meeting the requirements, which will generally involve some form of intumescent, mineral rope, or other flexible material as an incombustible insulant and barrier to flame and smoke. This again must not impair the performance of other joint components.

2.4.6 Appearance

The appearance of visible joints will have an effect on the overall appearance of a concrete structure, so attention should be given to integrating joints as part of the overall design. In a building or other structure this may involve architectural, or indeed, client input.

Joints cannot be disguised, so it is more realistic to express them, and to co-ordinate them with other visible surface features. Attention to detailing (Section 2.9), formwork (Section 3.2), and good concreting practice (Section 3.4) is essential if satisfactory appearance is to result[46].

2.4.7 Accessibility for inspection and maintenance

Any joint whose components cannot be relied upon for the full life of the building should be accessible for maintenance. In addition, joints should be accessible for regular inspection, allowing removal of detritus such as stones and other hard or injurious material that could impair performance of joints or components. This is essential.

All aspects of accessibility should be considered. It should be possible to reach joints using reasonable access equipment and without undue risk to operatives. On taller buildings abseiling may be a viable and economical alternative to cradles or major scaffolding. In addition, the joint design should allow removal of defective components, surface cleaning and preparation, and the replacement of components without risk or damage. This requires thought as to the width of joints and the depth at which components that may need replacement are placed.

2.5 PROVISION OF JOINTS

This section describes typical situations where movement joints are likely to be used. It is intended to be read with the relevant design code of practice or other references noted.

2.5.1 Building structures

Free movement joints in building superstructures will accommodate movements due to temperature variation and the effects of irreversible drying shrinkage.

The spacing of such joints is a matter for engineering judgement, which may be substantiated by calculations to demonstrate that the structure can accommodate the resulting forces induced by the sources of movement. Many UK designers will consider providing movement joints in large buildings at spacings of 60-70m[1]. It is recognised that the spacing may be increased if the structure is relatively flexible in response to horizontal movements, e.g. a flat slab roof on slender columns - although in such a case the secondary bending due to column sidesway would have to be designed for. Conversely, a smaller exposed structure with many stiff columns could benefit from a closer movement joint spacing to reduce bending and horizontal shear forces on the columns.

Uninsulated or lightly-insulated roof slabs, parapets, and exposed floor slabs may be jointed more closely (say at 20-40m), as these will react more rapidly to daily and seasonal temperature fluctuations.

Spacing of joints is clearly influenced by the range of temperatures experienced in service. The French concrete code[47] recommends provision of joints at spacings as close as 25m for structures in hot, dry, Mediterranean areas compared with 50m as the recommended spacing for cooler, damper regions. (Calculations are required if it is proposed to exceed these spacings.) A study by the US Building Research Advisory Board[48] suggested that, for framed symmetrical concrete structures, joint spacings up to as much as 180m might be feasible for heated buildings in zones with very modest temperature ranges.

Free movement joints are often desirable where concrete elements or building profiles change section abruptly (e.g. Figure 6) because the smaller cross-section is vulnerable to cracking from temperature variations.

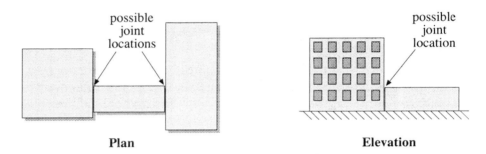

Figure 6 *Movement joints at change in building profile*

Free movement joints may be used to separate different structural systems within the same building. Figure 7 shows how a free movement joint will allow Block (b) to work as a frame under wind load (with larger sidesway deflections) while Block (a), with its stiff core wall structure, will deflect less. Without the joint, wind on the long face would produce torsional loading on the core and possibly troublesome distortional movements, as would occur with Block (c).

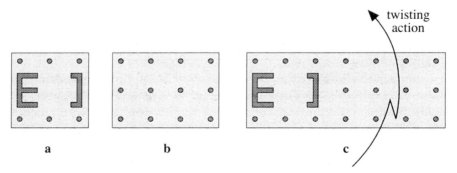

Figure 7 *Movement joints for unsymmetrical structures*

Free movement joints are commonly used against existing construction, especially where it is of a different structural form or not built of concrete. (If such a joint is not provided, the existing construction must of course have adequate strength to support additional loads from the new structure.)

Free movement joints, when used to allow horizontal movement, require that the structure must either be supported near the joints or be designed to cantilever. The former may sometimes give rise to double columns, often of reduced width so that the overall column width is retained for consistency (Figure 8). These narrower columns will be more flexible and hence more able to accommodate horizontal movements; but, if the original columns were slim, it may be impractical to build twin columns within the required space. Dowelled expansion or sliding joints may be an appropriate alternative.

Free movement joints are also provided to accommodate differential vertical movements where part of a structure is of substantially different height or weight than the remainder, or where the foundations for each part are of different types, in ground subject to settlement or consolidation (Figure 2). Mining subsidence may also call for movement joints to accommodate vertical and/or horizontal movements.

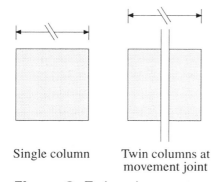

Single column   Twin columns at
                     movement joint

Figure 8 *Twin columns at a movement joint*

When such joints are provided to deal with foundation movement (essentially vertical, possibly rotational, sometimes horizontal), it is customary to continue the joint through the entire structure. Joints provided to accommodate temperature variation and irreversible drying shrinkage, however, are usually not continued below ground level. Temperature variations there are smaller, and shrinkage takes place more slowly (with more scope for creep relaxation) for elements in contact with soil. This may not apply to a multi-storey basement where joints might be continued down through the suspended slabs. Care is needed not to 'stop' such a joint within a continuous member (Figure 9) in view of the risk of unplanned cracking triggered by the 'notch' effect of the joint tip.

Where movement joints are being considered mainly to absorb irreversible drying shrinkage, it may be worth considering the alternative of leaving short bays unconcreted until later, and omitting movement joints. This may have implications, however, for structural stability where the floor plates act as horizontal girders, and for the contractor's programme.

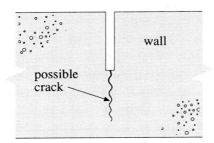

Figure 9 *Notch effect of interrupted movement joint*

2.5.2 Water-retaining structures

Guidance on the provision of joints is contained in BS8007, which offers three options to the designer for controlling early-age thermal contraction and irreversible restrained shrinkage. These are shown in Table 5, based on Table 5.1 of BS8007. Each option is accompanied in BS8007 by recommendations on the reinforcement to be provided to control cracking between joints.

Table 5 *Design options for water-retaining structures*

Option	Movement joint spacing
1 Continuous full restraint	No joints (but free movement joints at wide spacings may be desirable in walls and roofs that are not protected from solar heat gain or where the contained liquid is subjected to a substantial temperature range)
2 Semi-continuous: for partial restraint	a) Free contraction joints at ≤ 15m b) Alternate partial and free contraction joints (by interpolation) ≤ 11.25m c) Partial contraction joints ≤ 7.5m
3 Close movement joint spacing: for freedom of movement	Three sub-options as 2, but with joints at closer centres, calculated from Appendix A of BS8007 - using less reinforcement than option 2

2.5.3 Basements

Joints in basement construction are a prime source of leakage with the attendant issues of consequent damage and difficulty of remedial treatment. This is particularly true at movement joints, with the conflicting requirements of freedom to move and exclusion of water. Some form of flexible watertight barrier or "waterstop" is essential at such joints, but it is often judged preferable on balance to dispense with movement joints below ground (as discussed in Section 2.5.1) and concentrate on achieving monolithic water-excluding concrete construction[18].

BS8102 classifies levels of protection for basements according to use. For all except the first grade (car parking, plant rooms without electrical equipment, and workshops) one recommended form of construction is with reinforced concrete designed to BS8007, which would follow advice on joints given in Section 2.5.2

above. Other design approaches adopt BS8110 for the concrete design. These guidance sources are also relevant for ducts, access tunnels, and the like.

2.5.4 Ground-bearing floor slabs

The provision or the avoidance of joints in ground-bearing concrete floor slabs is generally recognised to be an important factor in their successful construction and performance. The risk of damage to any joints present will depend mainly on the nature and intensity of traffic. Guidance on the type and spacing of joints is given in Deacon[11]. Typical joints originating from Deacon are shown in Figure 10.

Joints can be eliminated if the ground slab is prestressed to overcome potentially damaging tensile stresses. Careful detailing at edges and columns is essential to allow free contraction of the entire slab, which must be constructed on a slip membrane. A Concrete Society Technical Report gives design guidance[49].

2.5.5 'Massive' structures

Much of the guidance in this report may be applicable to 'massive' concrete structures e.g. gravity dams. They are, however, sufficiently specialised not to be covered in detail here.

2.6 REINFORCEMENT

2.6.1 Use of reinforcement

The provision or omission of reinforcement across a movement joint is dictated by the function of the joint, as discussed in Section 2.3. Such reinforcement will either be bonded or debonded.

Bonded reinforcement is an essential feature of construction joints (see Section 4.6) where structural continuity is required and consequently the full area of longitudinal steel is carried across the joint.

Bonded reinforcement is employed in the tied partial contraction joint (see Section 2.3.3 and Figure 4(c)), using a reduced area of steel with the aim that - in particular - early-age thermal contraction and later drying shrinkage will be accommodated preferentially by movement at the joint, rather than by unplanned movement (i.e. cracking) elsewhere. The hinged joint (Section 2.3.5 and Figure 4(g)) is similarly intended to allow a component of free movement to occur at the joint (in this case rotation), while maintaining structural continuity to resist axial and shear forces.

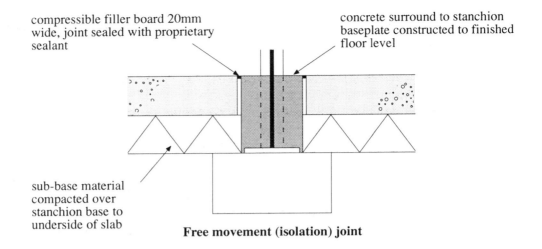

compressible filler board 20mm wide, joint sealed with proprietary sealant

concrete surround to stanchion baseplate constructed to finished floor level

sub-base material compacted over stanchion base to underside of slab

Free movement (isolation) joint

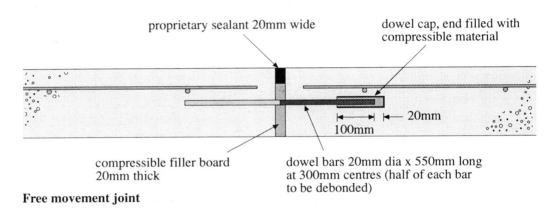

proprietary sealant 20mm wide

dowel cap, end filled with compressible material

20mm

100mm

compressible filler board 20mm thick

dowel bars 20mm dia x 550mm long at 300mm centres (half of each bar to be debonded)

Free movement joint

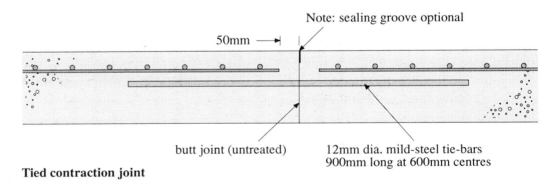

Note: sealing groove optional

50mm

butt joint (untreated)

12mm dia. mild-steel tie-bars 900mm long at 600mm centres

Tied contraction joint

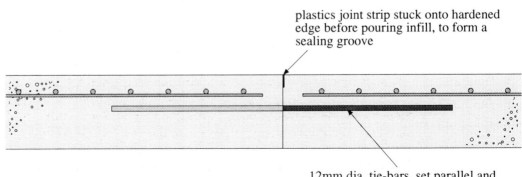

plastics joint strip stuck onto hardened edge before pouring infill, to form a sealing groove

12mm dia. tie-bars, set parallel and debonded over at least half length

Debonded contraction joint

Figure 10 *Typical joints in ground-bearing floor slabs*

Reinforcement that is debonded across a joint is intended to function by dowel action and to:

- allow free movement in a direction out of the plane of the joint (usually normal to it)
- transmit shear force across at least one axis in the plane of the joint.

Typical conditions where such behaviour is required include:

- at a debonded contraction joint in a ground-bearing slab (Figure 10)
- at an expansion joint between two phases of in-situ concrete construction, where shear loads have to be resisted without development of bending moments; two examples are a stair landing built between two previously-constructed stair walls (Figure 11(a)) - in this example free movement will be normal to the joint plane - and a walkway slab between two buildings (Figure 11(b)) in which one joint could have the requirement to accommodate horizontal movement along the joint axis.

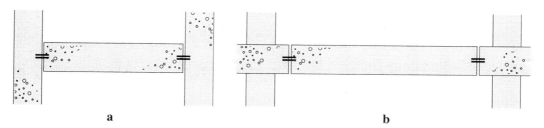

a   b

Figure 11 *Typical locations for debonded joints in superstructure*

There are two essential needs for reinforcement at such joints:

- it must be effectively debonded to allow free relative movement between the bars and concrete on at least one side of the joint
- the bars must be located consistently parallel and in the plane of the expected movement (normally also the plane of the member), and well-secured to avoid displacement during concreting.

2.6.2 Dowel bar design

For ground-bearing slabs the design of the reinforcement crossing expansion and contraction joints is covered in Deacon[11].

In lightly-loaded floors tied contraction joints may be formed by carrying through the normal slab reinforcement (usually fabric) but using only one layer, fixed near the bottom of the slab to allow drying shrinkage movement to occur in the top part of the slab (reducing the risk of uncontrolled surface cracking).

In more heavily-loaded floors, in thicker slabs and in bar-reinforced slabs, it will be more appropriate to provide dowel bars rather than fabric (which, with its cross-bars, will always be anchored on both sides of a joint and is therefore not suitable for joints intended to allow structural movement). Deacon gives empirical guidance on choosing bar diameter, spacing, and length.

In structural elements (other than ground-bearing floor slabs) the design of dowels to transfer shear forces requires calculation. Manufacturers of proprietary systems generally provide design tables backed up by a technical advisory service. For 'loose' bars the designer must refer to published guidance on the design of dowels. This is not covered in BS8110, but there is guidance in the 1990 edition of the CEB-FIP Model Code[32].

Guidance on friction values is limited. One manufacturer has indicated that friction forces may be based on a coefficient of friction for austenitic stainless steel in contact with itself, say a value of 1.0, quoted from a reference source[30]. The corresponding value for low-carbon normalised steel is 0.7-0.8. No test data specifically for dowel bars have been found.

The behaviour of dowels under shear is analogous to the behaviour of laterally-loaded piles in soil. Loading produces bending of the dowel and consequently bearing stresses on the surface zone of concrete. Failure is usually by yielding of the steel (particularly with smaller-diameter bars) or crushing of the concrete (particularly with larger-diameter bars). The Model Code gives a formula in clause 3.10 for calculating ultimate capacity, provided that the dowel has an embedment length of at least 8 diameters. The dowel must also be positioned with at least 8 diameters distance between its compressed face and the concrete surface to avoid premature fracture of the concrete by tensile cone failure in this area. There are also restrictions on minimum side distance (3 diameters) and edge distance from the bar face in tension (5 diameters).

The Model Code warns that the ultimate load shear displacement across a dowelled joint will be of the order of one-tenth of a bar diameter, e.g. 2mm for a 20mm bar. It will of course be less under service loading: with BS8110 partial load factors of 1.4 (dead) and 1.6 (imposed), the displacement under service load would be about 2mm divided by an average partial factor of 1.5, i.e. about 1.3mm. (This makes no allowance for non-linear deformation at higher loads and is therefore conservative.)

At least one proprietary dowel system uses pairs of dowels and tubes connected together by welded webs. These give increased load capacity and reduced shear displacement because of the greatly increased rigidity of the assembly.

Large local forces may be transmitted across such dowels, which must of course be safely resisted by the adjacent concrete. This will require a local design check with, often, an additional cage of enclosing reinforcement. U-bars should be used in thicker slabs and other members to inhibit tearing failure along the slab edge.

2.6.3 Durability

Reinforcement in gapped joints exposed to moisture is vulnerable to corrosion. *In situations where failure of such reinforcement would be unacceptable, it will be wise to use stainless steel or a suitable non-ferrous metal such as phosphor bronze.* This may also be appropriate in any exposed situation, since replacement of corroded reinforcement will inevitably be difficult and disruptive.

Coated carbon steel is not recommended in exposed situations as it is likely to shed its coating on the contact faces during repeated cyclic movements.

2.7 WATERSTOPS

Waterstops (also known as waterbars) are preformed strips intended to provide a physical barrier to water at joints, most commonly in basements and other below-ground construction.

Some form of waterstop is essential in movement joints required to exclude water. For construction joints there are arguments both for and against the use of waterstops, which leads many to recommend against their use, particularly when cast within the concrete section. As many of the issues are common to both movement and construction joints, it is appropriate to consider them here.

Arguments in favour of using waterstops are:

• they provide a physical barrier to water
• they remove complete dependence on the integrity of concrete at joints.

Arguments against their use are:

• they are vulnerable to damage during construction (but this may not become evident until later on, when location and repair of damage can be difficult)
• their presence impedes thorough preparation of joint surfaces before concreting, as tools cannot be used close to them for fear of damage
• compaction of concrete is more difficult around the waterstop profiles, leading to a risk of honeycombing and subsequent leakage through the concrete rather than through the joint
• they are liable to move or buckle during construction, again risking honeycombing and subsequent leakage, unless they are firmly secured.

There are many examples of the successful construction of basements and other below-ground concrete works both with and without waterstops, just as there are cases where leakage and other problems have occurred, again both with and without the use of waterstops. BS8007[4] states that "it is not necessary to incorporate waterstops in properly constructed construction joints" but also gives guidance on their proper use. BS8102 recommends their use in certain circumstances.

Consequently, it is not possible on the evidence categorically either to advocate or to disparage use of waterstops. Here, attention is focused on the key issues:

• practical design and sound detailing
• good workmanship, particularly in preparing concrete surfaces at construction joints, ensuring that concrete is thoroughly compacted and waterstops, if they are used, are secured in place
• competent and well-timed supervision.

Waterstops can be of 'passive' form in which a permanent physical water barrier is anchored in the concrete either side of a movement or construction joint, or of 'expanding' form in which a barrier activated by moisture is placed in a tied contraction or construction joint.

'Passive' waterstops

At present there is no British Standard for waterstops. These have commonly been made of rubber or polyvinyl chloride (PVC) in a variety of profiles (Figure 12). Galvanised steel and copper waterstops are now only used on rare occasions.

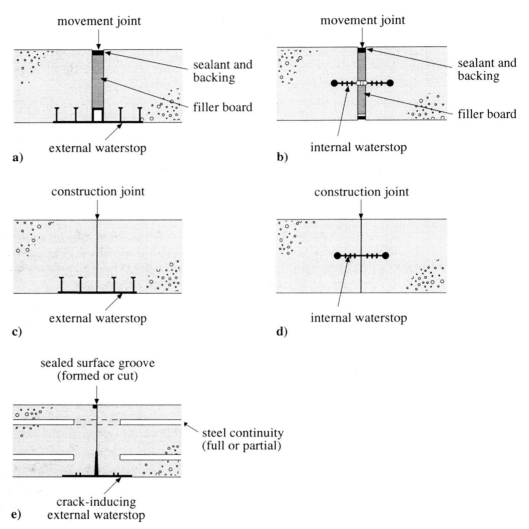

Figure 12 *Types of 'passive' waterstop*

There are two basic varieties of passive waterstops - those applied externally, and those applied internally (within the thickness of the concrete). External waterstops are only effective when placed on the face subject to a net clamping pressure, typically the outer face of a basement structure. Where this cannot be arranged, or water pressure can act either way, other approaches are required.

Internal waterstops can be effective against water pressure but the difficulties of properly incorporating them should not be underestimated. Improperly compacted concrete around the waterstop is a particular risk. *Indeed for horizontal construction, such as within slabs, the difficulties are so great that they should be avoided.*

Secure fixing is essential for waterstops to be effective. External waterstops are normally secured by nailing to the formwork. Internal waterstops are usually fixed by wiring to sufficiently robust sections of reinforcement.

Waterstops with central 'bulbs' or other voids (Types a and b in Figure 12) are intended for free movement joints, free or debonded contraction joints, and expansion joints where movement is expected. They may also be suitable for sliding joints provided that the waterstop material can accommodate the longitudinal sliding movement without tearing.

Simple 'passive' waterstops without central bulbs (Types c and d in Figure 12) are suitable only for use at tied contraction joints and at construction joints (see Section 4.7) where no significant movement can occur.

A third type is the crack-inducing waterstop (Type e in Figure 12) which can be used to encourage controlled cracking, especially in ground-bearing floor slabs.

In most 'passive' waterstop designs, the barrier element is manufactured with additional 'fins' on one or both sides, as appropriate, with the aim of increasing the path length for water entering behind the edge of the waterstop.

'Expanding' waterstops

More recently several new proprietary waterstops have appeared. They generally comprise small preformed strips of synthetic rubber incorporating a hydrophilic material that are secured against the face of a construction joint before the adjacent concrete is poured. Contact with water causes a swelling of the material, which is claimed to provide a compressive seal against further water ingress. Some initial leakage may occur as the process is not instantaneous and the material can be used only where it is initially and permanently fully contained, i.e. in tied contraction joints and construction joints. It is not suitable for use in other types of movement joint. Should water not be present continuously, the material will tend to contract, only to be effectively expanded by further water. The amount of swell may be affected by the water chemistry. There is, however, usually more than sufficient swelling to create a seal in the confines of the joint.

This form of waterstop offers good potential performance, especially in permanently damp or wet situations such as swimming pools. The ability to place them after the first pour of concrete makes them particularly attractive for linking new to existing construction. Figure 13 shows a diagrammatic view of an 'expanding' waterstop. Note that such a waterstop should be placed inside the reinforcement cage to protect concrete joint corners from spalling damage as the waterstop expands.

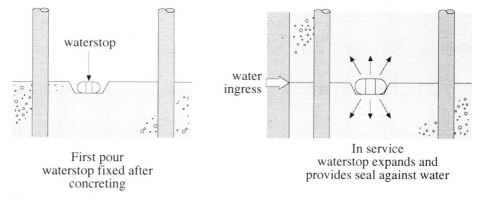

First pour
waterstop fixed after
concreting

In service
waterstop expands and
provides seal against water

Figure 13 *'Expanding' waterstop - typical detail*

Care in construction will be needed to avoid water coming into contact with the 'expanding' waterstop and causing it to expand prematurely, rendering it useless. (Sources of risk include rain.) The delay before expansion begins will afford some tolerance to dampness, but such waterstops should nevertheless be protected from water prior to concreting.

At least one of these materials has obtained an Agrément certificate relating to its use in concrete structures, and at least one is available in a formulation suitable for use in contact with potable water.

2.8 SEALANTS AND JOINT FILLERS

2.8.1 Sealants

The selection of sealants is covered in BS6213[23] which describes relevant joint design considerations (notably width), choice of sealant, design life, causes of failure, fire resistance, and maintenance. Types of sealant are described, and tables are provided to aid the designer in selecting suitable sealants for water-retaining structures, floors, building substructures and other water-excluding structures, roofs, external walls and cladding, windows and frames, internal finishes, and ceilings - all of which may involve sealing to concrete.

Further sources of advice are CIRIA Technical Notes 128[24] and 144[51] which deal with civil engineering sealants and the performance of sealant-concrete joints in wet conditions.

Significantly Technical Note 128 draws attention to the absence of tests to assess sealant resistance to sewage, seawater, or abrasion; it also notes the absence of tests in sealant standards to assess their effect on water quality, which limits the choice of suitable sealants to be used in contact with potable water. (Since this was published, work has started on a further CIRIA Project RP 472 investigating the microbiological deterioration of sealants under wet conditions.)

Key issues in selecting, specifying, and installing sealants are[23],[24],[33],[51]:

- identifying the essential considerations for satisfactory joint performance, particularly the assessment of causes, direction(s) and magnitude of movement to be accommodated
- selecting suitable sealants and compatible accessories such as joint fillers
- the crucial importance of achieving thorough adhesion to concrete surfaces by ensuring that the concrete is clean and, where appropriate, by using suitable primers
- choosing sealants not prone to the leaching-out of material that will stain adjacent concrete.

2.8.2 Joint fillers

Where movement is to be accommodated, the ideal joint filler is air. In practice the use of joint filling material is often desirable to prevent the joint space being filled by hard debris and the like, that could prevent free joint closure and cause damage to joint surfaces and waterstops. A filler may also be necessary to provide acoustic and thermal insulation and fire resistance, as discussed in Section 2.4, or backing to a weathertight sealant.

The criteria for selection of joint filler material are described in BS6093. Fillers should be robust, resilient, non-staining and should not extrude from the joint. They should also be compatible with any sealants used.

One particular point to note is that there is no such thing as a freely 'compressible' joint filler which can be compressed without the application of force. All solid materials require force to compress them, even cellular or foamed plastic compounds. It is usual to quote the compressive stress needed to achieve a 50% compression strain, i.e. a halving in thickness. BS6213 quotes typical values for common joint fillers (see Table 6).

The same document also notes that recovery after compression is usually less than 100%.

Table 6 *Compressive stress to produce 50% compression in common joint fillers*

Material	Stress to produce 50% compression (N/mm²)
Wood fibre/bitumen	0.7 - 5.2 (or more)
Bitumen/cork	0.7 - 5.2
Cork/resin	0.5 - 3.4
Cellular plastics and rubbers	0.07 - 0.34
Mineral or synthetic fibres	Dependent upon degree of compaction

2.9 DESIGN DETAILS

Joint details must take account of the following factors:

- the details must be simple and buildable, with tolerance between the components
- details must be robust (chamfers on joint corners are a useful way to reduce the risk of damage as formwork is removed)
- adequate cover must be provided to all reinforcement, with detailing to allow for the intrusion of waterstops (if present)
- joint profile must be wide and deep enough to accommodate the chosen sealant and its accessories (primer, backing strip, joint filler and debonding tape), and allow access to install the materials and subsequently to inspect and maintain them
- joint width must be matched to suit the movement range of the sealant to be used, both as built and at the extremes of anticipated movement (to avoid slumping or detachment of the sealant)
- the sealant must be specified (either by name or generic type) as part of the joint design, not left to a choice based on low cost or immediate availability
- where finishes are to be applied to the concrete surfaces, provision should be made for carrying the joint through the finish or other means for accommodating any movements.

3 Movement joints: site practice

3.1 REQUIREMENTS FOR FORMING JOINTS

The successful forming of movement joints involves:

- construction of concrete on both sides of the joint that is sound, well-compacted, and formed to the specified dimensions
- achievement of the specified joint gap dimensions
- reinforcement with adequate cover short of the joint, or continued across it with the necessary provision for anchorage or debonding to perform as specified, and securely held in place
- incorporation of waterstops and/or joint fillers where required
- subsequent installation of sealants where required.

In general respects the requirements for customary good practice should be followed. These are clearly defined in materials and workmanship standards for concrete construction such as BS8110: Part 1[2] and guidance documents such as CIRIA publications cited in this report, the Cement & Concrete Association (now British Cement Association) guide to concrete practice[25], and the Concrete Society guide on good practice in formwork[26]. Deacon[11] concentrates on concrete ground floors, but his practical guidance is of relevance to other elements, notably in the placing of reinforcement and the formation of joints.

Particular points relating to joints are discussed below.

3.2 FORMWORK

Normally the concrete on the two sides of the joint is poured on two separate occasions. This is not an inviolable rule, but special care is needed to secure any joint filler in place if concrete is placed on both sides simultaneously. However, it is desirable to carry enclosing formwork across the joint location to ensure that opposing joint faces are formed without misalignment. This ensures that sealants and fillers can be inserted without distortion while maintaining uniform thickness of these materials. Alignment of exposed faces also benefits appearance. Figure 14 illustrates the point.

Joint face formwork should be made accurately so that opposite face details (recesses, chamfers, etc) will match, for the reasons just given. It may be possible to make the joint face formwork material of the same thickness as the joint width in free movement and expansion joints.

The joint face formwork should have tapered sides (draw) sloped at about 1:8-10, wider at the surface, to allow easy removal of the form without risk of damage to the 'green' concrete arrises. As an alternative, a dimensionally stable but compressive material such as expanded polyethylene may be secured to the form face.

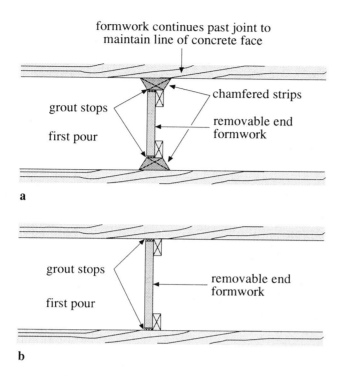

formwork continues past joint to
maintain line of concrete face

chamfered strips

grout stops

removable end
formwork

first pour

a

grout stops

removable end
formwork

first pour

b

Figure 14 *Continuity of formwork across a joint to minimise misalignment between joint faces*

3.3 REINFORCEMENT

3.3.1 General considerations

Reinforcement that is not to cross the joint must be fixed with adequate end cover to ensure durability. Reinforcement to be carried across the joint will be required by the design either to be bonded to the second pour of concrete (in tied contraction joints) or to be debonded (in expansion joints and debonded contraction joints).

In either case the reinforcement must be properly handled and protected, especially if it is to be exposed for a period of time, in which case corrosion protection may be needed[36]. Use of stainless steel, galvanised, or epoxy-coated bars may be warranted.

Reinforcement to be bonded to later-poured concrete is essentially in the same category as reinforcement at construction joints, and is accordingly covered in Section 5. Debonding is described below.

3.3.2 Debonding methods

Three methods are in principle available to achieve debonding:

• coating the bar with a debonding agent

- encasing the bar within a low-friction sheath (such as a plastic sleeve)
- fitting the bar into a rigid (usually metal) tube in which it is a sliding fit (the basis of several proprietary systems).

Since the objective is to minimise friction between the bar and the surrounding concrete, it follows that plain round bars stand the best chance of achieving this. Bars should be cut by sawing or other methods that leave a smooth steel surface. Shearing or cropping is unsuitable. Deformed and cold-worked bars will mechanically interlock with concrete if either an applied bonding agent or a low-friction sheath is used. There is little evidence on the performance of such bars within a rigid tube, but it may be surmised that the higher local contact pressures on the raised parts of such bars under shear load are more likely to 'dig' into the tube and cause locking-up than the lower, more evenly distributed contact pressures between two relatively smooth surfaces.

Coated debonding agents

Deacon[11] advises against the use of "any 'black' paint that is available", warning that "many bitumen-based paints actually increase bond". He recommends an "efficient" debonding compound consisting essentially of "66% of 200 pen bitumen blended hot with 14% light creosote oil and, when cold, brought to the consistency of paint by the addition of 20% solvent naphtha".

This conclusion was based on the results of a Cement & Concrete Association study[27]. A later study from the same source[28] again highlighted the indifferent performance of a number of debonding compounds. The Department of Transport (DoT) specification for highway works at that time countenanced use of debonding compounds, specifying that a slip of 0.25mm was to be achieved at a bond stress not exceeding 0.14N/mm^2. The tests showed that the compound prescribed at the time by the DoT and three other compounds (not identified, for reasons of commercial confidentiality) all complied with the DoT specification requirement, provided that concreting took place within a few days of application of the compound - the period varying between two and 14 days.

The study investigated the effect of delaying concreting up to 28 days after compound application, and found that this increased the bond stress at 0.25mm slip to between 0.18 and about 0.5N/mm^2.

The conclusions to be drawn from this are that debonding compounds should be applied no more than 48 hours before concreting, and that the structural design should take account of bond stresses of the order of 0.14N/mm^2.

No published data have been found on the effect on bond stress of alternating movements, repeated cycling, or degree of surface rusting of bars. Bitumen-based tapes are usually a very effective alternative coating agent to achieve debonding.

Low-friction sheaths

Deacon[11] recommends the use of plastic sleeves as an alternative to the suggested coated debonding agent described above. Tests carried out in the C&CA study[28] gave bond stresses in plastic sleeves of 0.03-0.17 N/mm^2 at 0.25mm slip. Plastic coated bars produced similar but slightly higher results at 0.25mm slip, with bond stresses of 0.10-0.15N/mm^2.

The report concluded favourably on the use of plastic sleeves, but recommended against using plastic-coated bars, as the bond stress achieved in these increased significantly prior to larger slippages.

The current DoT specification[29] prescribes the use of thin plastic sheathing (maximum thickness 1.25mm) for dowel bars in road pavement. No alternative is indicated. The requirement remains that a 0.25mm slip can be achieved at a bond stress not exceeding $0.14N/mm^2$.

The major practical advantage of plastic sleeves over coated debonding agents is clearly that the former can be used without undue concern about time. Plastic-sleeved bars can be prepared off-site in bulk, well in advance of construction.

Metal tubes

Several proprietary dowel-in-metal-tube shear connector systems are currently available in the UK. Figure 15 illustrates typical connectors schematically. Round bars are used at joints where the only movement to be accommodated is normal to the joint plane, whereas for joints where movement along the joint axis also has to be allowed for, either round or rectangular bars may be used, sliding in an oval or rectangular section sleeve.

The proprietary systems are usually made of stainless steel. This takes account of the exposure condition at moving joints, where water and other liquids may enter the joint and reach any reinforcement bridging it. The absence of significant corrosion with stainless steel will also minimise the risk of increased friction from corroding steel surfaces.

3.3.3 Alignment of dowel bars

It is intuitively obvious that if debonded dowel bars are not entirely parallel across a joint, then friction and applied forces will increase the resistance to movement across the joint and also apply additional stresses to the bars.

There is a lack of published data on the quantitative effects of such misalignment, but the current DoT specification lays down the following requirements for dowel bars in road pavements:

"Dowel bars shall be positioned at mid-depth from the surface level of the slab, ± 20mm. They shall be aligned parallel to the finished surface of the slab, to the centre line of the carriageway and to each other within the following tolerances:

"(i)   for bars supported on cradles prior to construction of the slab and inserted bars in two layer construction prior to placing the top layer:

    (a)   all bars in a joint shall be within ± 3mm per 300mm length of bar;
    (b)   two thirds of the bars shall be within ± 2mm per 300mm length of bar;
    (c)   no bar shall differ in alignment from an adjoining bar by more than 3mm per 300mm length of bar in either the horizontal or vertical plane;

(ii)   for all bars, after construction of the slab:

      (a)   twice the tolerances for alignment as in (i) above;
      (b)   equally positioned about the intended line of the joint within a tolerance
           of ± 25mm."

Deacon echoes this as-built alignment tolerance of ± 6mm per 300mm length of bar.

Proprietary systems may prescribe particular limits in trade literature, which should be adopted.

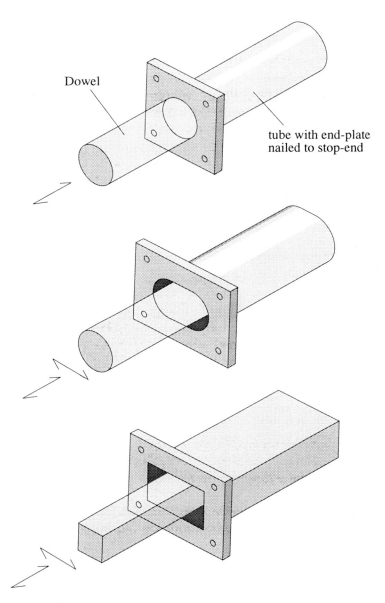

Dowel

tube with end-plate
nailed to stop-end

Figure 15 *Typical dowel-in-metal-tube shear connectors*

3.3.4 Provision for axial movement

Provision for axial movement of bars is necessary at expansion joints. The DoT specification requires that "a closely fitting cap 100mm long consisting of waterproofed cardboard or an approved synthetic material shall be placed over one end of each dowel bar. An expansion space 10mm greater than the thickness of the joint filler board shall be formed between the end of the cap and the end of the dowel bar."

Such movement provision is readily achievable with proprietary systems.

3.3.5 Support of dowel bars during construction

Clearly it is necessary to provide adequate support for the dowel bars so that they are both correctly aligned when fixed, and restrained robustly so that subsequent concrete placing and compaction does not alter their alignment.

For individual loose bars the method of support depends partly on whether the concrete is placed either side of the joint simultaneously or in two pours. In either case rigid support is essential.

Cradles, usually of steel, are commonly used to support each bar at two points near its ends. The British Reinforcement Manufacturers' Association has published drawing RMA/M1 (standard joint assemblies for contraction joints)[31], the essence of which is illustrated in Deacon[11]. A typical cradle is shown in Figure 16.

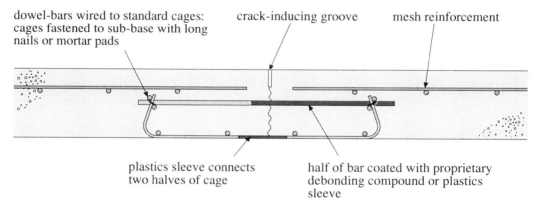

Figure 16 *Support cradle for dowel bars at contraction joint*

The plastic connecting sleeve would allow this cradle to be used in expansion joints (the cradle being fixed in two halves, one with each pour). However, this method does involve fixing and leaving cradles within the cover of one face (the bottom of a slab, for example), which may be unacceptable on grounds of both durability and appearance.

In exposed locations where appearance matters, it is possible (though expensive) to use stainless steel cradles. In such cases, and where appearance is important, a proprietary system is a practical and inconspicuous alternative - indeed, these systems are designed for expansion joints and sliding joints. These proprietary dowel systems usually have fixing plates or other devices designed for securing to the formwork stop-end, thus ensuring that the bars are fixed normal to its surface.

An alternative in visual concrete formed in a single pour is to support the cradles off the reinforcement cage. In any event it is wise to secure pre-formed cradles to adjacent reinforcement by tying wire to prevent them moving during concrete placing and compaction. A superior variation on this method is to use two stop-ends, drilled together, and placed a little way apart to provide support to the dowels. However, the success of this method is entirely dependent on the accuracy of drilling and placing the stop-ends.

The simplest but least satisfactory way to locate dowels between two pours is to push them through holes drilled in the stop-end. Without further securing (e.g. by tying or use of cradles) these will stand little chance of remaining in place correctly aligned throughout concreting. However, once restrained by the first pour, the projecting ends of dowels can be gently re-aligned if necessary before the second pour.

Where fabric is used as reinforcement across a tied contraction joint this can be fixed by conventional methods (mortar blocks, chairs, spacers, etc).

3.4 CONCRETING

3.4.1 General

Two particular concerns in concreting are to achieve good compaction at the joint face, especially around dowels, rebates, chamfers, etc; and to avoid displacing or damaging the various components such as waterstops, dowel bars and their supporting services. Both aims can be achieved by normal good practice[2],[11],[25], but there are particular points to be observed:

- concrete should preferably be moved into place at the joint indirectly from about 0.5m distance (e.g. using a vibrator) rather than being delivered directly from a skip or pump line. This is to minimise the risk of displacing the joint components, as noted above. (Falling or pumped concrete often has a substantial momentum)
- concrete placing should ensure that air is not trapped where it will result in voids, such as below a horizontally-fixed internal waterstop
- compaction tools should be used with care around the joint components to avoid displacing components.

3.4.2 Formation of surface grooves

Cracking is encouraged to occur preferentially at contraction joints (and not in the intervening concrete) by reducing the concrete section at these joints. Deacon[11] recommends use of a groove, formed or cut, of overall depth not less than one-quarter the member thickness. Such a groove acts as a stress-raising notch to encourage cracking there, as well as rendering the joint potentially weaker in tension through its reduced section.

Grooves may be formed in one of several ways:

- A crack-inducing strip of steel or plastic, may be fixed to formwork and/or inserted into the wet concrete on unformed faces such as the top of a ground slab. If strips are to be inserted into the wet concrete, it is important to do this while the concrete is still plastic. It is also essential to work the concrete back around the strips to eliminate voids and achieve a sound surface finish. This method is suitable where no formed joint is required, as in many ground slabs where a single pour may construct a number of bays at once[11] and where accuracy of line is not essential

- Timber fillets may alternatively be used on formwork where faces are accessible afterwards: fillets should have a 'draw' to assist removal of the formwork without damage. This method is appropriate for both formed and unformed joints

- Grooves can be cut by sawing once the concrete has hardened but before significant thermal contraction and shrinkage movements have begun. This should be within 8-72 hours of pouring. Use of a bottom crack-inducer can reduce the required depth of saw-cut but it can also increase the risk of uncontrolled cracking appearing on the top surface if sawing is not carried out soon enough. This method is suitable for unformed contraction joints particularly in ground slabs. Care is needed to locate the groove accurately. If it is not, the saw may damage top reinforcement; with a bottom crack-inducer there is a risk that surface cracking might appear close to the out-of-position groove. Premature sawing risks dislodging coarse aggregate, but left too long the sawing may not prevent early cracking.

3.5 WATERSTOPS

Waterstops must be accurately located, and securely held in place before and throughout concreting (see Section 5.4). A practical and effective method of achieving this should be agreed before work is carried out on site. Damage must be avoided if the waterstops are not to leak, and they may need to be protected prior to concreting. Steel fixing must avoid puncturing the waterstop material. Locating and remedying leaks are both troublesome tasks. A leak may manifest itself some distance away, as water can flow along the waterstop-concrete interface only to come through at a crack or an intersecting construction joint.

It is essential that individual sections of waterstop are properly joined to form a continuous barrier sealing all joints they cover. This requires the use of special junction pieces (shaped as ells, tees, etc) and the use of suitable adhesive or solvents to joint or 'weld' the sections and junction pieces together. Most waterstop manufacturers offer a technical advisory service to assist in specifying suitable profiles, and will also prepare detailed drawings of waterstop layout from which they prepare and supply the requisite material.

The more recent 'expanding' or hydrophilic waterstops are placed after the first concrete pour, being secured by adhesive. Most manufacturers recommend that they are set into a formed rebate in the concrete (Figure 12) to reduce the risk of damage from reinforcement fixing, concreting, and casual knocks.

In general these waterstops are at less risk of damage during construction than are the passive varieties.

3.6 JOINT FILLERS AND SEALANTS

Fillers are usually fixed after the first face of concrete is poured. Sealants are installed later as a finishing operation.

Joint fillers are normally in the form of a board. The material used should be:

- dimensionally stable
- inert
- resistant to rot (or impregnated to make it resistant)
- readily compressible (if to be used in expansion joints).

Various proprietary fillers are available. Most are plastic-based (expanded polyethylene, etc) or of fibreboard impregnated with bitumen.

It is essential that fillers are securely fixed in place before concreting.

Guidance on selection of sealants is given in BS6213[23] and BS6093[12]. CIRIA Technical Note 128[24] should be consulted for advice on sealants in wet conditions. CIRIA Special Publication 80[33] deals with sealant application practice.

3.7 MAINTENANCE

3.7.1 Concrete

In general, properly designed and constructed concrete at joints should not require maintenance.

One area that has traditionally given rise to concern (and indeed troublesome failures) is the jointing of ground-bearing slabs, particularly when trafficked by fork-lift trucks and other heavy wheeled vehicles. The continual stressing of sharp corners (Figure 17(a)) can cause spalling and lead to an untidy appearance. Ideally, joints should be narrow with a stiff sealant flush with the surface. Other approaches include chamfering the joint corners (Figure 17(b)), and use of steel corner angles cast-in (Figure 17(c)).

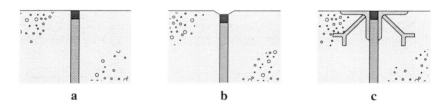

a   b   c

Figure 17 *Slab corner details with greater and lesser risk of damage*

3.7.2 Waterstops

In almost every case waterstops are inaccessible for maintenance without great disruption. This serves to emphasise the need for care in the selection, handling, and incorporation of waterstops in a structure.

3.7.3 Joint fillers and sealants

BS6093, BS6213 and CIRIA Special Publication 80 all give guidance on maintenance. Periodic inspection is recommended to identify deterioration or damage before it becomes more serious and more costly to repair. BS6213 gives the expected service life of sealants as between 10 and 20 years depending on the material, while BS6093 recommends that joints be inspected at intervals of one-fifth the expected service life, with an additional inspection after one year's service for joints subject to movement.

The expectation that all sealants will need replacement - and probably more than once - within the building lifespan should concentrate the designer's attention on ensuring accessibility to joints for inspection and sealant renewal.

4 Construction joints: design considerations

4.1 FUNCTIONS

Construction joints are a necessity of the building process, allowing construction of the concrete structure in manageable sections. This ensures that each session's pour can be supplied, placed, compacted, and finished within the time available.

Construction joints differ from movement joints in that no allowance is made for relative movement across the joint; indeed the objective is for the concrete structure to behave as if it were monolithic across the joint with no significant impairment of strength. This implies that both reinforcement and concrete must be effectively continuous across the joint, with particular implications for the lapping or joining of reinforcement and for the treatment of the existing concrete joint surface to produce the bonding or keying (if required) to the new concrete once it is placed.

So achievement of the required strength is always a major consideration for construction joints. Other issues of importance, depending on circumstances, may be:

- appearance
- durability and resistance to moisture
- location and spacing of construction joints.

4.2 STRENGTH REQUIREMENTS

A construction joint will normally be required to transmit one or more of the following types of stress:

- axial tension
- axial compression
- flexure (in or out of the plane of the member)
- shear
- torsion.

Of these, axial tension is in strength design always assumed to be resisted solely by the reinforcement. (In reality, of course, concrete has some tensile strength, as is recognised in prestressed concrete design and must be present to provide the diagonal tensile resistance to shear. This strength is also taken into account when considering the possibility of cracking, see Section 2.2.) The other types of stress all involve some force transfer across the concrete joint surfaces, either by compression or shear.

4.3 APPEARANCE

The appearance of construction joints is of importance in exposed concrete, and especially so when the concrete finish is of aesthetic importance. An untextured smooth concrete surface is particularly liable to be marred by a ragged construction joint.

A further consideration is that adjacent pours of concrete will almost invariably differ slightly in colour, reflecting (often minor) variations in the colour of cement and aggregate, in the mix proportions, in the amount of mixing and compaction, and in the formwork material, finish, curing, and any release agent or other coating it receives.

Both issues suggest that where appearance is a concern it may be preferable to 'express' the construction joint, for instance by forming it with, and at, a rebate. The evident line of the rebate will often effectively mask the actual concrete joint, while the latter - when in shadow as on horizontal joints - will be less conspicuous anyway. A method for achieving this is illustrated in Figure 18.

4.4 DURABILITY AND RESISTANCE TO MOISTURE

Properly formed, prepared, and constructed construction joints will generally afford durability comparable to that provided by the adjacent concrete. It may be argued that liquid moisture can be drawn through construction joints by capillary action if the joint is not 'perfectly' formed. Even if it is well formed some leakage may occur. It is often the practice to specify waterstops in water-resisting and water-retaining structures (see Section 2.7), although BS8007[4] states in clause 5.4 that "it is not necessary to incorporate waterstops in properly constructed construction joints". The most important factor is good workmanship and supervision in forming construction joints (see Section 5). The resistance to the passage of water can also be affected by the formwork used in the construction (see Section 5.2.2).

4.5 LOCATION AND SPACING OF CONSTRUCTION JOINTS

The main criteria for joint location and spacing are:

• the practical limits on what volume and/or area of concrete can be finished in one session
• the strength required and achievable at the joints
• the presence of restraints to contraction of the hardening concrete, affecting the possibility of early-age thermal cracking
• compatibility with appearance and jointing of applied finishes.

In recent years deep concrete lifts and large volume pours have become more common as part of 'fast-track' construction. These are dealt with in a recent CIRIA study[34] which describes the planning, organisation, and practical actions needed for success.

Construction joints are typically located at the junction between horizontal elements (beams, slabs, stair landings, etc) and vertical or inclined elements (columns, walls, stair flights, etc). They are also used to divide up large lengths or areas of individual elements. In the first case the spacing of joints is determined by member dimensions. In the second it is often based on the practical considerations noted above.

BS8110: Part 1[2] recommends that:

"The number of construction joints should be kept to the minimum necessary for the execution of the work. Their location should be carefully considered and agreed before concrete is placed. They should normally be at right angles to the direction of the member."

In vertical and inclined members a further consideration is the lateral pressure exerted by the unset concrete. CIRIA Report 108[35] provides a method for calculating pressure on formwork. Two examples may be cited. High walls and columns may be cheaper to build if poured in more than one lift, in view of the substantial and more costly falsework needed to resist the high lateral pressures that a single pour would generate against formwork in the lower parts of the element. Secondly, it is common practice with an 'infill' building to have walls or columns tight against the adjacent flank walls. These new concrete elements may be specified to have a joint filler board providing separation from the existing walls (to allow free relative movement). However, it is not then possible to place formwork behind the new element. If concrete is poured in too high a lift there is a very real danger of the adjacent wall being cracked, or even collapsing. The risk is increased if the wall is in poor condition and inadequately bonded or tied to adjacent construction.

It is often recommended (for example in the C&CA guide to concrete practice[25]) that construction joints in beams and slabs should be located at about one-third of the way along the span. For continuous members this should be a location where bending moments are low and, under normal loading, the shear force is modest. This will not always be the case, for example in a cantilever or indeed a simply-supported beam or slab; in both cases there is a substantial bending moment and shear at the third point. It may be preferable to do without a construction joint, but if one is provided particular care will be needed in its preparation to achieve good flexural and shear capacity across the joint. Current thinking is that beam and slab construction joints should be located at supports to simplify construction, although thought must be given to the strength requirements of the joint (see Section 5.2.).

Construction joints at the top of vertical or inclined elements should be at or just above the underside of the supported members. This will make cleaning out and inspection of formwork easier, once this has been fixed. Joints at the foot of vertical inclined elements are usually formed at the surface level or just above, forming a 'kicker' of some 75-200mm height. This kicker acts as a locating device for the next lift of formwork, but it must be formed soundly if it is not to prove a weak point in terms of strength and water-exclusion. 'Kickerless' construction is an alternative (see Section 5.5.3).

Most joints in ground-bearing slabs are detailed and specified to allow movement (see Section 2.3), except for prestressed slabs where no such provision is necessary. The location of construction joints in such slabs will be determined by practicality.

In water-resisting and water-retaining structures designed to BS8007[4] joints are usually detailed to allow some movement (see Section 2.5.2). The spacing of construction joints in structures designed to CP110 (and, by extension, to BS8110) is recommended in the CIRIA guide to waterproof basements[18] as follows:

• walls - not greater than 5m
• slabs - not greater than 10m in the longer direction

although the guide does note that these spacings may be increased without risk provided careful attention is given to mix design and construction procedures. Early-age thermal cracking is identified as a particular hazard to be borne in mind.

4.6 REINFORCEMENT

It is common to achieve continuity of reinforcement across construction joints using projecting starter bars for lapping with subsequent steel, but other methods are available and these are discussed in Section 5.3.

Protection of reinforcement at construction joints is covered in a recent CIRIA study[36].

4.7 WATERSTOPS

The design issues relating to the use, or not, of passive waterstops, are discussed in Section 2.7, whilst the construction practice is discussed in Sections 3.5 and 5.4. Most of the issues are common to both movement joints and construction joints.

4.8 SEALANTS

Sealants are sometimes used at construction joints to protect arrises in ground slabs, and as a second line of defence in watertight construction (see Section 3.6).

4.9 DESIGN DETAILS

Provision of rebates to mask construction joints has been discussed in Section 4.3. The detail may be used where the concrete is to be exposed in use, as in Figure 18.

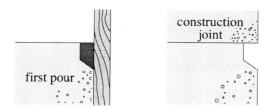

Figure 18 *Forming a rebate at a construction joint*

5 Construction joints: site practice

Recommendations for the forming and preparation of vertical and horizontal construction joints are currently given in a number of authoritative sources, including BS8110: Part 1[2], BS8007[4], Deacon[11] for ground-bearing concrete floor slabs, and good practice guides from the British Cement Association[25],[52] and the Concrete Society[44]. Their general recommendations should be followed in regard to formwork, reinforcement and concreting. Specific issues for construction joints are considered below, supplementing the guidance for movement joints in Section 3.

5.1 PRACTICAL REQUIREMENTS

The design requirements for a construction joint in structural concrete have already been identified in Sections 4.2-4.4 as:

- adequate strength
- acceptable appearance
- adequate durability and resistance to moisture.

The contractor has to achieve the above as well as constructional accuracy, but will also be concerned with cost and speed of building, and the effective use of labour.

The two stages in producing a vertical construction joint are firstly to provide a former against which the flowing concrete is stopped until it has hardened; and secondly to prepare the hardened concrete surface as necessary to achieve the required continuity with the subsequent concrete pour.

Horizontal construction joints are not formed, but occur at the free surface of poured concrete. The problems of forming a stop-end do not arise. On the other hand, some surplus water inevitably rises to the top of the pour along with finer cement and sand particles which are deposited on top of the true concrete. When this water has evaporated, a layer of so-called laitance remains. This is weaker than the concrete in tension, compression, and shear; is not watertight; and is not a satisfactory bonding material. Generally it is preferable that laitance should be removed (see Section 5.7).

5.2 EFFECT OF FORMING AND PREPARATION METHODS ON CONSTRUCTION JOINT STRENGTH

5.2.1 Available information

A review of current practice on the treatment of construction joints, and of available research information dating back to 1930, was carried out by Brook at the Cement and Concrete Association. This was published in 1969 as CIRIA Report 16[41]. A complementary C&CA Technical Report (TRA 414)[42] appeared in the same year, with more detailed information on previous research results.

This study concluded that further research was needed on a number of aspects. Of particular note was the recommendation that attention should be given to the condition of construction joints in existing structures to identify how they were performing. (Unfortunately this does not appear to have been followed up in any published study.) Brook noted that an unbonded construction joint becomes a contraction joint, but took a sanguine view of this. "If the joint does not have to prevent water leakage, there may be many situations in which it is better for a shrinkage crack to develop at a vertical joint than elsewhere. It can be said that concrete is bound to contract and what better place than ... where the shrinkage crack is controlled along a predetermined straight line. The alternative may be random cracking elsewhere ..." This philosophy is appropriate where strength requirements, watertightness, and durability considerations do not demand a bonded joint.

This study also recommended "a close look at the condition of construction joints made with joggles". A joggle, known in the USA as a key or keyway, is a formed trapezoidal rebate which provides mechanical resistance to shear across the joint (see Figure 19 and Section 5.5.1).

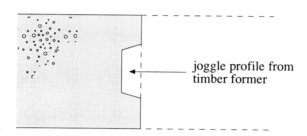

joggle profile from timber former

Figure 19 *A joggle joint*

Tests on construction joints were carried out at the C&CA by Monks and Sadgrove and published as Technical Report 42.483 in December 1973[40]. These investigated the effect of construction joints, with both 'smooth' and 'roughened' surfaces, on the flexural and shear performance of reinforced concrete test beams. These were of practical size (150mm wide, 280mm deep) as compared with several earlier investigations in which flexural and tensile tests were made on small-section unreinforced concrete specimens. The tests were able to take account of the contribution to shear resistance of the effects of aggregate interlock and dowel forces in supplementing the 'pure' frictional resistance across the compression zone of the beams. This work also formed the basis of an article by Monks on the treatment of construction joints[43]. The findings of these studies are summarised in Sections 5.2.2 - 5.2.5 below.

Since then there appears to have been little work done to examine the effects on strength of the treatment of construction joints. There have been some developments in materials and techniques for forming construction joints (see Section 5.5), but relatively little has been published on their effects on joint strength.

5.2.2 Tensile strength

Investigators have usually tested the concrete-to-concrete 'bond' in plain (i.e. unreinforced) concrete loaded in tension or flexure across a construction joint. Such tests necessarily take no account of reinforcement as a means of carrying tensile forces.

It was concluded by Brook in CIRIA Report 16 that "the highest values of bond across a joint are achieved by casting against a clean, slightly roughened joint face which is dry and which is not treated with a priming layer of mortar".

5.2.3 Compressive strength at a vertical construction joint

For a vertical construction joint, i.e. a joint involving formed faces, it can be concluded from the above studies and tests of beam behaviour that the presence of the joint has no significant effect on the compressive strength or stiffness of the jointed concrete element. (This of course assumes complete removal of any deleterious or weak materials at the joint face, and also assumes effective compaction of concrete at the joint. The same assumptions apply to other conclusions.)

5.2.4 Flexural and shear strength at a vertical construction joint

The presence of a construction joint in a reinforced concrete section subject only to flexure (i.e. with minimal shear) does not significantly affect the stiffness and ultimate flexural capacity of the section. This is the case even when bond is prevented across the joint, or when the abutting joint faces are smooth and untreated. Cracking is likely at such a joint; but where the reinforcement is designed to current code guidance on serviceability and crack control, such cracking at serviceability loading is unlikely to exceed 0.3mm in width[40].

A similar section with a construction joint subject to bending and shear will have much the same flexural and shear capacity as a monolithic section, provided the joint surface is effectively roughened. The same applies to a section subject to high shear and little or no bending.

Sections in which the construction joint surface is smooth and untreated will fail at substantially lower shear loads (typically 40% lower) than would a corresponding monolithic or roughened-joint section. This emphasises the importance of achieving a roughened surface in joints subject to significant shear forces.

5.2.5 Compressive, flexural, and shear strength at a horizontal construction joint

There is little experimental evidence of the effect of surface laitance on compressive, flexural, and shear strength in horizontal construction joints. However, its known relative weakness suggests that sections incorporating joints with a layer of laitance would not achieve the same strengths as sections with roughened joints. There appears to be no evidence to justify abandoning the general practice of removing laitance before casting fresh concrete, except perhaps where design forces are very modest.

BS8110 gives design guidance on the horizontal shear strength of concrete in such locations, for example at the interface of an in-situ topping intended to work compositely with precast concrete slabs. Another common example is the slab forming the flange of a T-beam, poured against the previously-cast beam web. A brushed, screeded, or rough-tamped surface is rated up to 50% stronger than an untreated surface, while one that has been washed to remove laitance, or treated with retarder and cleaned, is rated up to 75% stronger than an untreated surface. The presence of 'nominal' links across the interface increases the ultimate horizontal shear strength typically by a factor of three, but where the horizontal shear stress exceeds these strengths then all the horizontal shear force is to be carried by designed reinforcement crossing the joint and properly anchored on either side. (Nominal links are those with a cross-section area of at least 0.15% of the contact area.)

5.3 REINFORCEMENT

5.3.1 Projecting reinforcement

Projecting starter bars are 'traditional'. They offer the advantages of simplicity and cheapness, with the disadvantages that formwork is expensive, bars 'get in the way', and are fairly easily damaged, particularly in ground works where projecting vertical bars (e.g. from piles) are at risk from site vehicles. Once accidentally bent or snapped off they may be difficult to repair.

Many engineers specify mild steel for projecting bars from piles because this is a more ductile material, and can be rebent with a greater chance of success, than the stronger hot-rolled and cold-worked high yield bars.

Projecting bars must be properly supported, particularly during concreting. Vertical bars of any diameter are liable to need additional restraint; the smaller diameter bars are likely to buckle, while larger bars, having greater rigidity, may move unnoticed in the formwork if accidentally struck on the projecting portion. (This could reduce cover and lead to durability and appearance problems later.)

Pre-drilled stop-ends provide a convenient template for fixing projecting reinforcement across a vertical construction joint. The reinforcement should then be stabilised against movement during concreting by wiring to adjacent bars and/or with supports beyond the stop-end.

Vertical projecting reinforcement at horizontal construction joints must likewise be stabilised. The cost and difficulty of doing this (and also of stabilising the formwork) may be a decisive factor in the choice of one or several pours for a tall column or wall (coupled with concrete pressures on formwork, see Section 4.5 above).

A further important consideration is that of safety. Projecting horizontal reinforcement is particularly hazardous if it projects beyond formwork anywhere near eye level. So also are vertical starter bars at ground level. In such cases the bar ends should be covered with blunt caps, which are often of plastic.

5.3.2 Bent-up reinforcement

One alternative to projecting starter bars is reinforcement bent up within the formwork. One arm of each bar (typically bent as an L-bar) is located in contact with the formwork; after concrete has been placed and the formwork struck, the bars can be pulled out and straightened (Figure 20). This involves rebending of the bars (consequently small-diameter bars of mild steel are preferable) with the attendant risk of local damage to the concrete where the bars are embedded. For highway and bridge works such rebending must be approved by the Engineer[29]. To minimise the risk of brittle fracture, rebending should not be carried out below 5°C.

Figure 20 *Bent-up reinforcement*

Heating to a cherry red colour (800-850°C) may be used to assist bending - but if applied to cold-worked steel this will cause reversion to mild steel strength. Heating should be applied over the full length of the bend so that the bar can be straightened out and not left with a crank.

The traditional way of straightening such bent bars is with a scaffold tube, which increases leverage. But this approach can cause spalling of concrete around the bar if applied too vigorously. Rebending should be done around a former of suitably large radius (see BS 8110: Part 1[2]). After straightening-out the bar should be allowed to cool naturally. Quenching may embrittle the steel.

Care is needed to straighten fully embedded bars to their correct position without introducing unacceptable kinking. CIRIA Report 92[37] reviews reinforcement connector and anchorage methods, including cast-in bent bars. It recommends that the arm to be bent out should be enclosed in a box-out to aid subsequent bending. (The box-out should embrace the bent part of the bar, so that when straightening takes place the concrete behind the bend is not spalled off.) It also advises that such bars should be of mild steel, and not exceed 16mm in diameter.

Bending-out should be left as late as possible so that concrete has gained sufficient strength to minimise the risk of cracking or spalling during the process. A minimum of 48 hours is recommended.

5.3.3 Stop-ends with integral reinforcement

These use bent-up bars as described in Section 5.3.2 but, instead of being individually fixed as loose bars, they are purpose-designed and are supplied as a group already embedded in a stop-end, which simplifies fixing. The stop-end may be temporary and made of plastic foam or steel sheet, or permanent and formed in precast concrete. Several proprietary systems are available, some having full CARES approval. Bars used are either ordinary reinforcement complying with BS4449[38], or surface-hardened bars with improved ductility which also meet the requirements of BS4449. Some systems offer only a small choice of bar sizes and spacings, which may limit their usefulness in particular situations where heavier reinforcement or closer bar spacing is necessary.

5.3.4 Reinforcement couplers

A detailed review of reinforcement couplers appears in CIRIA Report 92[37]. Couplers involve use in the first stage concrete either of projecting bars or recessed threaded sleeves. To these, further reinforcement is subsequently connected by one of several methods including sleeves (swaged, wedged, or bolted) and threading. Potential problems with projecting reinforcement are as described above; those for threaded sleeves are as outlined below for cast-in sockets. Close supervision is essential.

5.3.5 Cast-in sockets

A variety of cast-in sockets are available into which threaded reinforcing bars may subsequently be screwed. Even if the open socket end is tight against the formwork, there is a risk of grout seeping into it, thereby blocking the threads and making later screwing impossible. Open ends must accordingly be temporarily blocked off before concreting. Accurate positioning is important.

A similar risk arises if the uncapped socket is exposed for any time, especially in an aggressive atmosphere that would accelerate rusting of the threads. Some form of capping is therefore essential.

It is tempting, where rusting is a concern, to use galvanised or sherardised sockets. Where sufficient thickness of zinc is applied to the steel surface, long-term protection against rusting can be achieved. Some fixings and sockets may only have thin electroplated coatings of zinc or other metals. The protection against rusting from these systems will be more limited, and may only be sufficient to prevent corrosion during storage.

5.3.6 Welding reinforcement

Hot-rolled reinforcement may be welded to achieve continuity across construction joints. Cold-worked reinforcement may also be welded, but the heating involved is likely to reduce its strength towards that of the steel prior to cold-working. Welding allows reinforcement to be connected within a shorter length than is needed for laps in concrete.

BS7123[50] gives guidance on procedures for welding. The essentials for successful welding are clean materials, competent workmanship, and close supervision.

5.4 WATERSTOPS

If waterstops are used they will influence choice of joint forming and preparation methods. The presence of a waterstop may make some joint-forming methods impractical. An internal waterstop, for example, prevents the proper installation of expanded metal as a stop-end (see Section 5.5.2).

The risk of puncturing, or otherwise damaging, the waterstops should be considered when choosing joint preparation methods. Abrasive or power-tool techniques would certainly be unsuitable. Air or water jetting would be least likely to result in damage (see Table 7 and Section 5.7).

When used, waterstops must be properly joined to form a continuous barrier and must be securely fixed in position so that they are not displaced before or during concreting. If not properly secured, the waterstops may slump into the wet concrete, leaving voids which greatly increase the risk of leakage.

External waterstops are usually spiked to formwork near the edge of the waterstop material. Internal waterstops are often wired to the reinforcement cage through purpose-designed eyelets placed near the edge of the waterstop material. The manufacturer's guidance on fixing methods should be followed whenever available.

5.5 FORMWORK FOR CONSTRUCTION JOINTS

5.5.1 Vertical construction joints

Vertical construction joints should be formed against a stop-end, which should essentially be at right angles to the member axis. Angling-back the joint face in beams and slabs by, typically, 10mm in 300mm is believed by some engineers to give improved shear transfer. This is intuitively obviously where the angling-back induces some compression across the joint, although unsupported by tests (Figure 21). A stop-end should be of simple construction to allow easy installation and removal, although the presence of reinforcement continuing across the joint does complicate this. Stop-ends - other than in expanded metal - should be sealed to prevent excessive grout loss at the joint (as should all formwork joints). Grout loss leads to honeycombing and reduced strength, as well as being unsightly. Foam plastic strip is a widely-used, cheap, effective grout seal.

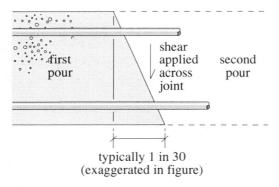

Figure 21 *Angled face on vertical construction joint*

For exposed concrete surfaces the forming of a rebate is a useful way of making construction joints less conspicuous (see Section 4.3). This may readily be done by forming a chamfered fillet on one or both faces of the formwork, as in Figure 18.

'Joggle' joints were commonly used in the past, particularly in watertight construction and where significant shear forces were present, as at the foot of a cantilever retaining wall. The joggle was usually formed by fixing a tapered block to the stop-end in the middle of the member (see Figure 19). The joggle can usefully increase the length of the path to be followed by leaking water and can also improve shear resistance. Joggled joints are used less often nowadays, mainly because of the additional expense of forming them. They are, however, still used for joints in radiation-shielding vessels, and sometimes in large-scale mass concrete work. For most concrete construction roughening of the joint - necessarily involving removal of any surface material - will increase the likelihood of producing a leak-free joint with good shear resistance.

A surface retarder may be applied to aid subsequent surface removal (this may be subject to the designer's agreement). It should be used according to the manufacturer's instructions. Care should be taken to avoid reinforcement being contaminated with mould oil, any retarder, and other applied formwork treatments, as these will adversely affect bond[36]. Retarder should not be applied too soon before concreting, as it may be washed off by subsequent rainfall or hosing-down.

The bottom of a vertical joint is at particular risk of collecting debris, before and during the clean-out of formwork prior to concreting. Debris can include lumps of timber and tying wire as well as laitance, concrete fragments, and dust from the joint surface treatment. These items can adversely affect joint strength and impair appearance, so should be removed. Specific attention to this is needed, especially at the base of a column or wall lift where access is difficult. One solution is to remove the stop-end and clear the debris just before concreting. A suction device may be effective on shallower lifts. A common solution uses small slots made in the formwork (where the concrete face is not to be exposed in use), plugged after cleaning-out. In any event it is worth cleaning-out before side forms are fixed, to reduce the amount of debris to be dealt with later.

A CIRIA study[39] was carried out in 1976 with the aim of identifying and evaluating ways to improve the forming of stop-ends (and, by extension, vertical construction joints). It listed the most serious practical problems at the time as:

• cutting and shaping timber (then and still the commonest material for forming stop-ends) - a slow and laborious task
• increasing costs of timber
• the high probability of damaging stop-ends when removing them from around reinforcement after concreting, restricting re-use of stop-ends (or, alternatively, the problems of grout loss if the stop-ends are loosely fitted round steel)
• the general necessity to scabble or otherwise prepare the formed concrete surface before casting the second pour.

The hope was that a simple, better, cheaper way might be evolved of forming stop-ends in vertical construction joints. This need was expressed clearly in this CIRIA study, but unfortunately no radical new solution has emerged as yet that meets the criteria of simplicity, improvement, and cheapness. There have, however, been some developments.

The available methods are now considered.

5.5.2 Stop-end formers

Common stop-end formers are:

• timber
• expanded metal
• ribbed matting
• pre-formed continuity strips incorporating reinforcement (see Section 5.3)
• precast concrete

- expanded plastic sheet (e.g. polyethylene)
- reinforced plastic sheet.

Alternatives identified in the CIRIA study and considered worth testing were:

- timber and compressible strip combinations
- 'socks' of plastic sleeving with granular fill
- inflated deformable tubes
- fabric stop-ends.

Timber

Timber, including faced timber sheeting, is still the commonest material for vertical construction joint stop-ends. There remains the slow laborious task of notching it to fit around reinforcement (especially in beams, and in slabs and walls with two layers of reinforcement), and the likelihood of damage during recovery.

Timber stop-ends generally result in a 'smooth' face to the concrete, although treatment with a retarder will facilitate subsequent roughening of the concrete surface.

Expanded metal

Expanded metal mesh - usually of galvanised steel - is a viable alternative to timber, although careful workmanship in placing and compacting concrete is necessary for good performance. Excessive vibration during concreting may drive excess cement paste and fines through the mesh. This may make joint preparation harder, and may produce honeycombing. Too little vibration is more of a problem as it may not adequately compact the concrete. Achieving the optimum degree of compaction is essential for critical elements such as in water-retaining construction.

Proprietary expanded metal mesh is commonly available in two forms. The first is plain flat sheet, which requires a firm supporting stop-end to prevent it distorting when concrete is placed. The second form incorporates raised ribs. These stiffen the mesh in one direction, so that it needs additional support only from discrete posts (usually of timber) rather than from continuous backing. The raised ribs are placed into the area to be poured first, so that they do not foul the supports. They therefore bond into the first pour, making mesh removal impractical.

The principal makers of expanded metal mesh publish technical advice covering its use as stop-ends. Such advice should be followed. It usually recommends that fresh concrete should not be placed directly against the mesh; instead it should be deposited about 0.5 metre away from the stop-end and moved into place with vibrators or other compacting equipment. These should not be allowed to touch the mesh, to avoid mechanical damage and excessive bleeding-through of cement paste and fines, and should at their closest be about 0.3 metre away. Vibration should be continued until the cement paste protrudes through the mesh perforations.

Plain flat mesh can either be removed or left in place. Removal soon after the concrete has hardened will leave a roughened profiled surface which needs, at most, only light brushing to expose the coarse aggregate. If the mesh is left in place there should be no need to prepare the concrete face except in the case of watertight construction where the need to remove all laitance calls for surface treatment or, more practically, removal of the mesh. A further concern for water-retaining structures is the ability of the mesh to provide a preferential waterpath through the concrete section. Where a high level of confidence in water resistance is required, the use of left in place expanded metal mesh might be considered questionable.

If mesh is to be left in place, for reasons of durability it should not extend into the prescribed cover zone to avoid rusting, staining, and possible spalling. (Stainless steel mesh of course is not affected by this need, but is more expensive.)

The practicality of placing reinforcement needs to be considered when using mesh stop-ends. The reinforcement is usually fixed first. Mesh can then be locally cut or deformed to fit around the bars, and timber strips placed within the cover zones. This is cheaper and quicker than notching the timber around bars, and also eases removal of timber after concreting.

Ribbed matting

Ribbed matting of rubber or plastic may be used in much the same way as expanded metal to achieve a roughened joint surface. A rib depth of about 3mm is recommended[40]. The matting must be removed when the concrete has hardened, and needs timber or other backing to stop it distorting unduly under the pressure of the wet concrete. As with expanded metal, it is easiest to use if confined within the reinforcement and used in conjunction with timber edge strips.

Pre-formed continuity strips incorporating reinforcement

These are discussed in Section 5.3. Temporary strips of plastic foam, steel sheet, etc may be removed after the concrete has hardened to allow surface preparation; permanent strips of precast concrete are considered below.

Precast concrete

Precast concrete stop-ends can be effective and save labour, especially if detailed to incorporate lapping reinforcement to link adjacent pours. Rebating both stop-end faces to form joggles can eliminate the need for surface preparation. Alternatively a roughened surface finish can be incorporated.

Expanded plastic sheet

Expanded polystyrene and polyethylene sheet can be cheaply cut and notched on site to form a full-profile stop-end, although these materials will need backing to prevent them bulging or fracturing under wet concrete pressure in view of their low stiffness and modulus of rupture. Some adhesion to hardened concrete may occur when the sheet is pulled away. The concrete surface must then be roughened as necessary.

Reinforced plastic sheet

Plastic sheet reinforced with an integral metal mesh can form a cheap but effective stop-end. Reinforcement is pushed through the plastic. The concrete surface should be prepared as necessary after the sheeting has been pulled away.

Other methods

Of the other methods tested in the CIRIA study[39], none appears in the subsequent 17 years since publication to have received much attention. Only one, the inflatable or pre-formed tube, has found some use in stop-ends in watertight or water-retaining construction. Inserted in the centre of a formed construction joint, it results in a continuous void which may be grouted at a subsequent stage in construction (or later, should leakage be experienced through the joint). A more recent version uses a steel-reinforced semi-permeable flexible synthetic tube which can be secured to the stop-end within a steel mesh cage. After concrete has hardened, the tube is injected with resin: again, this can be done either as a pre-emptive measure, or subsequently if found necessary.

5.5.3 Horizontal construction joints

Horizontal construction joints do not need formwork to produce the joint face.

Kickers are often used to 'start off' columns and walls from beams and slabs. These are formed in a first pour in ordinary concrete construction, as it is accepted that it is easier to position accurately a shallow timber box than the full height of formwork. Kickers must be treated as discrete structural members when pouring, and not regarded as somewhere to dispose of indifferent concrete. The proper concrete mix must be used, with compaction and curing of the same standard as used elsewhere.

In watertight construction using waterstops, kickers may offer the benefit of raising the construction joint clear of the slab top surface. This has the advantage of lifting the waterstops clear of slab reinforcement - particularly important with centrally-placed wall base waterstops, which might otherwise foul the bars. In such cases the kicker must be poured integrally with the member below.

Minimum kicker heights should be:

- 75mm if the kicker is poured integrally with the member below (but higher if necessary with centrally-placed waterstops to avoid fouling reinforcement in the member below)
- 150mm if the kicker is to be poured separately from the member below (to allow use of a vibrator within the plastic concrete).

An integral kicker must be used where a visible horizontal construction joint is to be masked behind finishes but these do not extend 150mm above floor or roof level. Where the finish is asphalt and a horizontal tuck is required, this may be formed at the top of the kicker with a detail similar to that in Figure 19.

Kickers more than 200mm high are seldom used.

Once constructed the kicker forms a convenient locating device for the next lift of formwork.

The alternative of 'kickerless' construction[45] removes one stage of the construction process and can therefore speed up operations. One way of achieving it is to secure a rectangle of timber strips to the horizontal concrete beneath with shot-fired pins, after marking out the element outline accurately. The strips are fixed beyond this outline by the thickness of the vertical formwork, which can then be slotted into place (see

Figure 22). Another method of fixing the formwork is provided by cast-in T-bars, against which the forms can be gently tightened to give a more secure fixing at slightly greater cost[45].

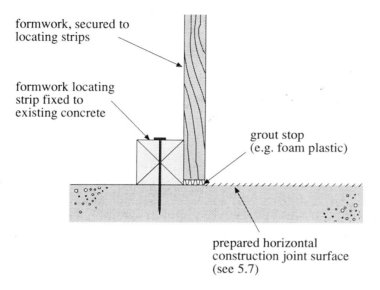

Figure 22 *Timber formwork locating strips for kickerless construction*

Cast-in fixings are another way to secure battens and vertical formwork. It is important with kickerless construction both to anchor the formwork securely, and to seal the gap under the formwork (e.g. with foam strip) to minimise grout loss and reduce the risk of honeycombing.

Horizontal joints in exposed concrete surfaces may be neatly concealed using fillets on one or both shutter faces, as discussed above for vertical joints and as shown in Figure 18.

'Joggled' joints are not commonly used today in horizontal construction joints, for the reasons discussed for vertical joints in Section 5.5.1.

Horizontal joints must be cleaned out before concreting, as described in Section 5.5.1.

5.6 CONCRETING

5.6.1 Vertical construction joints

The joint surface should be prepared if necessary, as described in Section 5.7. The surface should be clean and dry or 'saturated surface dry', i.e. wet but with no free water present. It should not be primed with a coat of mortar or grout: this does not improve bond, may impair appearance on exposed concrete surfaces, and is an additional and unnecessary task[25].

Concrete should be placed no closer than 0.15-0.3m from the joint, and then vibrated to move it into place against the stop-end. This is intended to aid good compaction, and will also reduce the risk of impact damage from concrete falling onto the stop-end and any waterstop.

With expanded metal stop-ends it is important to ensure that concrete is well compacted at the joint, but the compacting tool should not dwell overlong at the stop-end, nor should it be allowed to touch the expanded metal, as excessive loss of cement paste and fines may occur through the metal perforations. Less energy is reflected back into the pour when the metal is not backed with timber, so more compaction than usual may be needed. However, in this case the far side of the metal is exposed, and compaction can thus be halted before too much paste and fines has 'bled' through the holes. (The lost material will of course harden, and have to be broken out before the next pour.)

5.6.2 Horizontal construction joints

The joint surface should be prepared if necessary, as described in Section 5.7. The surface should be clean and dry or saturated surface dry (see Section 5.6.1). It should not be primed with a layer of mortar or grout.

The initial discharge from a ready-mixed truck may be 'bony' (i.e. short on fines). This should not be placed on the joint surface, where it is particularly important to have well-proportioned readily compactable concrete.

Concrete should be spread over the joint surface to a depth of about 0.3m[25] and compacted thoroughly at intervals not exceeding 0.5m, to minimise the risk of honeycombing at the bottom of the lift. Further concrete should then be placed, keeping the compacting tool(s) submerged so that displaced air can move fairly freely up from the compacted concrete through uncompacted material. Concrete should be brought up in even lifts.

In very narrow members it may be helpful to use an 'elephant's trunk' piece of ducting to ensure that concrete fills the formwork evenly and to avoid splash marking of formwork to exposed surfaces[34].

5.7 PREPARATION OF JOINT SURFACES

Surface treatment of the hardened concrete joint surface is carried out for the following purposes:

- to remove laitance - particularly when the joint is horizontal - where the concrete has to carry direct or flexural compression or shear across the joint
- to remove laitance in all joints, vertical and horizontal, where the concrete itself is intended to exclude or retain water and other aqueous liquids
- to ensure good bond across the joint where the concrete has to transfer tension and/or significant shear
- to clean and prepare the existing concrete surface so that it has no adverse effect on the fresh concrete, for example by impeding its hardening (due to contaminants - particularly mould oil) or by drawing in water (with the risk of honeycombing).

In general, therefore, the joint surface must be clean and (usually) free of laitance.

Where good bond is necessary the following are required:

- removal of surface laitance, cement paste and fines (typically to 2-3mm depth)

- exposure of the coarse aggregate (again to 2-3mm depth, without undercutting the matrix around it)
- minimal disturbance (ideally none) of underlying material.

The various treatment methods in current use and their suitability for particular situations are reviewed in Table 7. Other preparatory work is dealt with in Table 8.

For preparing the concrete joint surface, early treatment while the concrete is still 'green' is cheaper and requires less effort than later treatment when the concrete is harder. Early treatment should employ 'gentle' methods to minimise the risk of loosening coarse aggregate and otherwise weakening the joint; later on a more vigorous treatment, although necessary, must be applied with care to safeguard arrises and adjacent concrete faces from damage.

Table 7 *Treatment methods for surface at construction joints*

Method for removal of laitance and roughening of concrete surface	Use on vertical joint faces (with stop-end removed typically 4-6 hours after concreting)*	Use on horizontal joint faces*
Gentle air-jetting (air line must have filter to exclude risk of oil being deposited on concrete)	Suitable for use within 2-4 hours of concreting: not always practical unless concrete hardens quickly and stop-end removed promptly; low risk of damage	Within 2-4 hours of concreting; low risk of damage
Brushing with choice of soft and stiff (but not wire) brushes, aided by a water spray at low pressure - not a water jet	Apply on removal of stop-end (concrete may then be too hard to treat by this method unless retarder used); not always practical if waterstops present	Within 4 hours of concreting; not always practical if waterstops present
Gentle water jetting with low pressure water flow directed onto joint face	As above; usable if waterstops present; practical up to about 6 hours after concreting	As for vertical joints
Wire brushing with water washing/gentle jetting	Practical up to about 24 hours after concreting; not always practical if waterstops present	As for vertical joints
Application of needle gun or hand-held percussive tool (scaling hammer, etc)	Best left until concrete at least 3 days old, to avoid risk of loosening coarse aggregate arrises; care needed to avoid damage to waterstops	As for vertical joints
Wet abrasive blasting using grit or sand (dry blasting demands extensive additional health precautions)	As for needle gun/tool treatment; not really practical on thin sections; messy	As for vertical joints
'Hacking' by hand or hand power tool	**Not recommended** as it can loosen embedded coarse aggregate and damage concrete arrises; thorough joint preparation unlikely to be achieved	**Not recommended,** as for vertical joints

Note

*   Times will vary, being shorter with higher site temperatures, richer mixes, and lower water-cement ratios

Table 8 *Preparatory work at construction joints prior to concreting*

Work	Vertical joint faces	Horizontal joint faces
Cleaning out formwork - essential	Debris must be removed from bottom of shutter, see Section 5.5.1	As for vertical joints, see also Section 5.5.3
Wetting of joint surface before concreting	Not essential, but may be only way of ensuring clean joint face - must drain free water away before concreting	As for vertical joints
Application of priming coat of mortar or grout	Not recommended	Not recommended
Sealing of formwork joints	Essential	Essential
Application of surface retarder at joint face to delay surface set of first concrete pour(†)	Acceptable, will allow more flexibility in delaying treatment described above	As for vertical joints, but careful control needed to avoid retarder affecting set of underlying concrete; well-proportioned mix essential to avoid bleed-water diluting retarder and reducing its effectiveness

Note

(†)   Effects of retarder in delaying set are best estimated from retarder manufacturer's data

5.8 SEALANTS

CIRIA Special Publication 80[33] gives guidance on sealant application practice.

Key issues are:

- concrete preparation to achieve a clean surface of the required texture
- the surface to be dry
- appropriate primer to be applied
- sealant application to follow correct procedure.

The manufacturer's recommendations must always be followed.

5.9 MAINTENANCE

Properly designed and constructed concrete at construction joints should not require maintenance.

Guidance on inspection and maintenance of sealants is summarised in Section 3.7.3 above.

6 Conclusions and recommendations

6.1 MOVEMENT JOINTS

Movement joints afford a means of accommodating movements, and accordingly of reducing the risk of structural failure or damage due to stresses induced in the concrete if movement is restrained. The principal sources of movement that should be taken into account when considering the need for movement joints are:

- temperature variation
- early-age thermal movement as concrete cools after hydration
- irreversible drying shrinkage
- ground movement.

Various forms of movement joint may be used, depending on the translational and rotational movements to be permitted:

- the free movement (or isolation) joint
- the free contraction joint
- the partial contraction joint (either tied or debonded)
- the expansion joint (either free or dowelled)
- the hinged joint
- the sliding joint or bearing
- the seismic joint.

In designing such joints it is essential to consider non-structural functional requirements, including:

- durability
- resistance to water passage
- sound insulation
- thermal insulation
- resistance to fire
- appearance
- accessibility for inspection and maintenance.

Movement joints may be provided in building superstructures at intervals ranging typically between 25m and 180m, depending on the magnitude of anticipated movements and the stiffness and disposition of vertical structures (particularly shear walls). Joints at relatively close centres are desirable for stiff structures subject to relatively large movements, while a wider spacing will be more appropriate for flexible structures (e.g. frames) subject to smaller movements. Closer movement joint spacing may be appropriate for water-retaining structures and ground-bearing floor slabs. It is often preferable to omit movement joints in basement structures, relying instead on reinforcement to control cracking.

Reinforcement to transmit forces across movement joints must also be designed, detailed, and constructed to allow free movement and to avoid corrosion damage, as appropriate.

Waterstops are often used to exclude water at movement joints. They may be externally applied (on the face subject to water pressure) or internally within the concrete. Their use calls for particular attention to sound practical detailing for 'buildability', thorough compaction of concrete to obviate voids or honeycombing, and timely supervision as - once concreted in - a damaged waterstop or voided concrete cannot be easily or reliably repaired to eliminate leaks.

Movement joints should be sealed and/or filled as appropriate to exclude hard or deleterious materials (including water) that may affect the satisfactory performance of the joint.

6.2 CONSTRUCTION JOINTS

Construction joints are a necessity in any practical structure to allow it to be concreted in more than one operation.

Such a joint should not significantly affect:

• structural strength
• appearance
• durability and resistance to moisture.

In practice the presence of a junction between two pours will have an effect. The existing concrete face may have to be treated to remove surface laitance. This will usually be necessary to ensure adequate compressive, shear or torsional strength across the joint. Laitance removal is essential in 'watertight' construction. The almost inevitable (if only slight) visual difference in the colour of exposed concrete placed either side of a construction joint, and the risk of unsightly joint lines and/or grout leakage, can be minimised by:

• use of a rebate, with the construction joint concealed within the 'shadow zone' on its inner face
• formwork design to ensure alignment either side of the joint (best done by carrying formwork across the joint)
• effective joint sealing before concreting
• thorough compaction.

The spacing of construction joints will generally be determined by:

• practical limits on placing, compacting, and finishing concrete in one session
• presence of restraints to contraction of the hardening concrete.

In general concrete work it is preferable to have as few construction joints as practical. In 'watertight' basement construction, such joints are more closely spaced, typically no further apart than 5m centres in walls and 10m centres in floors.

Reinforcement may be carried across a construction joint as projecting bars; bent-up bars fixed as part of the reinforcement cage that can be bent out later (taking care to avoid mechanical damage to either steel or adjacent concrete); bars set in proprietary stop-ends (often of concrete), again to be bent out later; as couplers connecting to bars on either side of the joints; or as cast-in sockets to receive threaded bars. A further possibility is to weld projecting reinforcement where this is shorter than needed to achieve full lapping.

Opinions differ on whether waterstops are essential and effective. Where waterstops are not used, the watertightness of the joint is entirely dependent on a well-prepared sound joint surface using the procedures identified above. Great care is needed where waterstops are used, to avoid damaging them and to achieve thoroughly-compacted concrete.

Various materials are used to form stop-ends for construction joints. Timber is traditional, but other materials include:

• expanded metal (sometimes left in place)
• ribbed matting
• pre-formed continuity strips incorporating reinforcement
• precast concrete
• expanded plastic sheet
• reinforced plastic sheet.

In horizontal construction joints, a kicker (if used) should be at least 75mm high (if poured integrally with the member below) or 150mm (if poured separately).

At the time of concreting, a construction joint face should be clean and preferably dry. (Wetting may be necessary to ensure cleanness of formwork and of the joint face, but free water should be drained out.) Priming with mortar or grout has been shown not to improve bond, and it may impair appearance on exposed concrete faces. It is not recommended.

At vertical joints, concrete should be placed 0.15-0.3m from the stop-end and then moved into place by compaction; this reduces the risk of impact damage to the stop-end and any waterstop. Particular care is needed to achieve good compaction at expanded metal stop-ends.

At horizontal joints, concrete should be placed in a first pour of about 0.3m depth and compacted thoroughly before further concrete is placed. Sheet baffles may be provided against formwork for exposed faces and drawn up as concreting proceeds, to reduce the risk of splash-marking. Use of an 'elephant's trunk' duct is, however, usually a simpler and more effective alternative and is particularly useful in narrow members.

Joint surface preparation will usually be necessary for the reasons already noted. The aims should be to:

• remove laitance and cement paste and fines (typically to 2-3mm depth)
• expose (but not disturb) coarse aggregate, again to 2-3mm depth
• expose coarse aggregate without disturbing the underlying material.

Early, 'gentle' treatment will involve less effort as the concrete will not be so hard at this time. This also minimises the risk of damage from more 'aggressive' methods.

'Gentle' methods include air-jetting; brushing (not with a wire brush) accompanied by a low-pressure water spray; and gentle water jetting. More 'aggressive' methods include wire brushing with water washing; use of a needle gun or scaling hammer; and wet abrasive blasting. 'Hacking' is not recommended as it is unlikely to achieve thorough joint preparation. It also risks loosening embedded coarse aggregate and damaging concrete arrises.

References

1. ALEXANDER, S.J. and LAWSON, R.M.
    (1981)
    Design for movement in buildings
    CIRIA Technical Note 107

2. BRITISH STANDARDS INSTITUTION
    Structural use of concrete
    Part 1. Code of practice for design and construction
    BS 8110: 1985

3. BRITISH STANDARDS INSTITUTION
    Structural use of concrete
    Part 2. Code of practice for special circumstances
    BS8110: 1985

4. BRITISH STANDARDS INSTITUTION
    Code of practice for design of concrete structures for retaining aqueous liquids
    BS 8007: 1989

5. HARRISON, T.A.
    (1981)
    Early-age thermal crack control in concrete
    CIRIA Report 91

6. BUILDING RESEARCH ESTABLISHMENT
    (1979)
    Estimation of thermal and moisture movements and stresses: Parts 1 to 3
    BRE Digests 227, 228, and 229

7. BRITISH STANDARDS INSTITUTION
    Code of practice for use of masonry
    Part 3. Materials and components, design and workmanship
    BS 5628: 1985

8. INSTITUTION OF STRUCTURAL ENGINEERS
    (1989)
    Soil-structure interaction: the real behaviour of structures
    ISE

9. BIRT, J.C.
    (1974)
    Large concrete pours - a survey of current practice
    CIRIA Report 49 (see also Ref.[34])

10. THE CONCRETE SOCIETY
    (1982)
    Non-structural cracks in concrete
    Concrete Society Technical Report No. 22

11. DEACON, R.C.
(1986)
Concrete ground floors: their design, construction and finish
Cement and Concrete Association (now British Cement Association)

12. BRITISH STANDARDS INSTITUTION
Code of practice for design of joints and jointing in building construction
BS 6093: 1993

13. BRITISH STANDARDS INSTITUTION
Durability of buildings and building elements, products and components
BS 7543: 1992

14. SIMPSON, B. et al.
(1989)
The engineering implications of rising groundwater levels in the deep aquifer below London
CIRIA Special Publication 69

14a. KNIPE, C.V., LLOYD, J.W., LERNER, D.N. and GRESWELL, R.
(1993)
Rising groundwater levels in Birmingham and the engineering implications
CIRIA Special Publication 92

15. HER MAJESTY'S STATIONERY OFFICE
(1991)
The Building Regulations 1991
HMSO Statutory Instruments 1991 No. 2768

16. DEPARTMENT OF THE ENVIRONMENT
(1991)
The Building Regulations 1991
Approved Document C: site preparation and resistance to moisture
HMSO

17. BRITISH STANDARDS INSTITUTION
Code of practice for protection of structures against water from the ground
BS 8102: 1990

18. CONSTRUCTION INDUSTRY RESEARCH AND INFORMATION ASSOCIATION
(1978)
Guide to the design of waterproof basements
CIRIA Guide 5
NOTE: Water Resisting Basement Construction - A Guide published 1995.
CIRIA Report 139

19. BRITISH STANDARDS INSTITUTION
Code of practice for sound insulation and noise reduction for buildings
BS 8233: 1987

20. DEPARTMENT OF THE ENVIRONMENT
    (1989)
    The Building Regulations 1985
    Approved Document L: conservation of fuel and power
    HMSO

21. BUILDING RESEARCH ESTABLISHMENT
    (1989)
    Thermal insulation: avoiding risks
    BRE Report 143/HMSO

22. DEPARTMENT OF THE ENVIRONMENT
    (1991)
    The Building Regulations 1991
    Approved Document B: fire safety
    HMSO

23. BRITISH STANDARDS INSTITUTION
    Guide to selection of constructional sealants
    BS 6213: 1982

24. CONSTRUCTION INDUSTRY RESEARCH AND INFORMATION
    ASSOCIATION/WATER RESEARCH CENTRE
    (1987)
    Civil engineering sealants in wet conditions
    CIRIA Technical Note 128

25. BLACKLEDGE, G.F.
    (1987)
    Concrete practice
    Cement and Concrete Association (now British Cement Association)

26. THE CONCRETE SOCIETY
    (1986)
    Formwork: a guide to good practice
    The Concrete Society

27. WEAVER, J.
    (1967)
    A comparison of some bond-preventing compounds for dowel-bars in concrete
    road and airfield joints
    Cement and Concrete Association Technical Report TRA 403

28. MULAJKAR, V.N.
    (1980)
    The effectiveness of debonding materials for dowel bars
    Cement and Concrete Association Technical Report 537

29. DEPARTMENT OF TRANSPORT et al.
    Specification for Highway Works
    HMSO

30. BRANDES, E.A. and BROOK, G. (eds.)
    (1992)
    Smithells' metals reference book
    Butterworth-Heinmann

31. BRITISH REINFORCEMENT MANUFACTURERS' ASSOCIATION
    n.d.
    Standard joint assemblies for contraction joints
    BRMA Drawing RMA/M1

32. COMITÉ EURO-INTERNATIONAL DU BÉTON
    (1991)
    CEB-FIP Model Code 1990: final draft
    CEB Bulletins d'Information 203 to 205

33. CONSTRUCTION INDUSTRY RESEARCH AND INFORMATION
    ASSOCIATION/BRITISH ADHESIVES AND SEALANTS ASSOCIATION
    (1991)
    Manual of good practice in sealant application
    CIRIA Special Publication 80/BASA

34. BAMFORTH, P.B. and PRICE, W.F.
    (1995)
    Concreting deep lifts and large volume pours
    CIRIA Report 135

35. CLEAR, C.A. and HARRISON, T.A.
    (1985)
    Concrete pressure on formwork
    CIRIA Report 108

36. BUSSELL, M.N. and CATHER, R.
    (1995)
    Care and treatment of steel reinforcement and the protection of starter bars
    CIRIA Report 147

37. PATERSON, W.S. and RAVENHILL, K.R.
    (1981)
    Reinforcement connector and anchorage methods
    CIRIA Report 92

38. BRITISH STANDARDS INSTITUTION
    Specification for carbon steel bars for the reinforcement of concrete
    BS 4449: 1988

39. KIRK, G. and VENABLES, R.K.
    (1976)
    Development of improved stop-ends: report on preliminary testing and scheme for further work
    CIRIA Technical Note 67

40. MONKS, W.L. and SADGROVE, B.M.
    (1973)
    The effect of construction joints on the performance of reinforced concrete beams
    Cement and Concrete Association Technical Report 42.483

41. BROOK, K.M.
    (1969)
    Treatment of concrete construction joints
    CIRIA Report 16

42. BROOK, K.M.
    (1969)
    Construction joints in concrete
    Cement and Concrete Association Technical Report TRA 414

43. MONKS, W.L.
    Treatment of construction joints
    Concrete February 1974
    Vol: 8 (No: 2), 28 to 29

44. THE CONCRETE SOCIETY
    (1988)
    Joints in in-situ concrete
    Concrete Society Digest No. 10

45. BENNETT, D.F.H.
    (1988)
    Kickerless construction
    British Cement Association Publication 47.023

46. MONKS, W.
    (1988)
    Appearance matters - 1
    Visual concrete: design and production
    British Cement Association

47. ASSOCIATION FRANCAISE DE NORMALISATION
    (1986)
    Technical rules for the design and calculation of reinforced concrete structures and buildings using the limit states method (English translation of French original document BAEL 83)
    AFNOR Bulletin No.62, Part 1, Section 1

48. BUILDING RESEARCH ADVISORY BOARD
    (1974)
    Expansion joints in buildings
    Federal Construction Council Technical Report No. 65
    National Academy of Sciences

49. THE CONCRETE SOCIETY
    (1984)
    Post-tensioned flat-slab design handbook
    Concrete Society Technical Report No. 25

50. BRITISH STANDARDS INSTITUTION
    Specification for metal arc welding of steel for concrete reinforcement
    BS7123: 1989

51. CONSTRUCTION INDUSTRY RESEARCH AND INFORMATION
    ASSOCIATION
    (1993)
    Performance of sealant-concrete joints in wet conditions
    CIRIA Technical Note 144

52. BRITISH CEMENT ASSOCIATION
    (1993)
    Concrete on site: 7
    Construction joints
    British Cement Association Publication 45. 207

53. THE CONCRETE SOCIETY
    (1988)
    Concrete industrial ground floors
    Concrete Society Technical Report No. 34

Bibliography

AMERICAN CONCRETE INSTITUTE
(1992)
ACI manual of concrete inspection (8th edition)
ACI

AMERICAN CONCRETE INSTITUTE
(1989)
Guide for concrete floor and slab construction
(ACI 302.1R-89)
ACI

AMERICAN CONCRETE INSTITUTE
(1989)
Building code requirements for reinforced concrete (ACI 318-89) and Commentary
(ACI 318R-89)
ACI

AMERICAN CONCRETE INSTITUTE
(1989)
Guide for measuring, mixing, transporting and placing concrete (ACI 304R-89)
ACI

BRITISH STANDARDS INSTITUTION
Steel, concrete and composite bridges
Part 7. Specification for materials and workmanship, concrete, reinforcement and prestressing tendons
BS 5400: 1978

BRITISH STANDARDS INSTITUTION
Steel, concrete and composite bridges
Part 8. Recommendations for materials and workmanship, concrete, reinforcement and prestressing tendons
BS 5400: 1978

BRITISH STANDARDS INSTITUTION
Workmanship on building sites
Part 2. Code of practice for concrete work
Section 2.2 Sitework with in-situ and precast concrete
BS 8000: 1990

BRITISH STANDARDS INSTITUTION
Eurocode 2: Design of concrete structures
Part 1. General rules and rules for buildings
DD ENV 1992-1-1: 1992

CONCRETE REINFORCING STEEL INSTITUTE
(1965)
Monolithic reinforced concrete: the CRSI manual of good practice
CRSI

CONSTRUCTION INDUSTRY RESEARCH AND INFORMATION ASSOCIATION
(1984)
CIRIA Guide to concrete construction in the Gulf region
CIRIA Special Publication 31

DEPARTMENT OF TRANSPORT et al.
Notes for (1991) guidance on the Specification for Highway Works
HMSO

GARBER, G.
(1991)
Design and construction of concrete floors
Edward Arnold

METZGER, S.N.
Better industrial floors through better joints
Concrete Construction August 1988
Vol. 33 (No. 8), 749 to 754

MONKS, W.L.
(1972)
The performance of waterstops in movement joints
Cement and Concrete Association Technical Report 42.475

NATIONAL BUILDING SPECIFICATION LTD.
(current edition)
The NBS Specification

ROWE, R.E. et al.
(1987)
Handbook to British Standard BS8110: 1985, structural use of concrete
Palladian Publications Ltd

SADGROVE, B.M.
(1974)
Water retention tests of horizontal joints in thick-walled reinforced concrete structures
Cement and Concrete Association Technical Report 42.498

SUPRENANT, B.A.
Construction joints for multistorey structures (where to locate them, how to form them)
Concrete Construction June 1988
Vol. 33 (No. 6), 577 to 581

SUPRENANT, B.A.
Shear keys for basement walls: pros and cons
Concrete Construction July 1987
Vol. 32 (No. 7), 620 to 624

U.S. DEPARTMENT OF THE INTERIOR, WATER AND POWER RESOURCES
SERVICE
(1981)
Concrete manual: water resources technical publication
US Department of the Interior

Summary

This report gives advice on the provision, specification, and construction of joints in new in-situ concrete construction. It covers both movement joints, which are intended to relieve potentially damaging stresses or strains, and construction joints, which are necessitated by practical limitations on the amount of concrete that can be poured, compacted, and finished in one operation.

Design considerations are reviewed, and guidance is provided on design and construction issues associated with joints. Of particular importance are the successful exclusion of moisture where necessary; the filling and sealing of joints; and the preparation of construction joint surfaces to ensure satisfactory performance. References and a bibliography are included, listing relevant UK sources and a number from overseas.

Bussell, M.N. and Cather, R.
Design and Construction of Joints in Concrete Structures
Construction Industry Research and Information Association
Report 146, 1995

CIRIA ISBN	086017 429 8
Thomas Telford ISBN	0 7277 2092 9
ISSN	0305 408 X

© CIRIA 1995

Keywords	Reader interest	Classification	
Concrete structures, construction joints, contraction joints, formwork, hinged joints, joints, movement joints, reinforcement, sealants, seismic joints, sliding joints, stopends, waterstops.	Design, specification, construction and supervising engineers involved in concrete construction. Equipment suppliers.	AVAILABILITY	Unrestricted
		CONTENTS	State of art report
		STATUS	Committee guided
		USER	Professionals concerned with concrete construction

Health and Safety

Construction activities, particularly on building sites, have significant Health and Safety implications. These can be the result of the activities themselves, or can arise from the nature of the materials and chemicals used in construction. The report does not endeavour to give comprehensive coverage of the Health and Safety issues relevant to the subject it covers, although specific points to note are mentioned where appropriate in the text. Readers should consult other specific published guidance relating to Health and Safety in construction.

Report 146   1995

Design and construction of joints in concrete structures

MN Bussell
R Cather

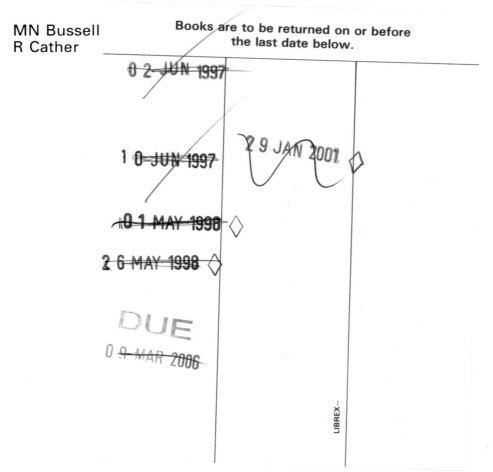

CIRIA

CONSTRUCTION INDUSTRY RESEARCH AND INFORMATION ASSOCIATION
6 Storey's Gate, Westminster, London SW1P 3AU
E-mail switchboard@ciria.org.uk
Tel 0171-222 8891   Fax 0171-222 1708